An Atlas of Fungal Ultrastructure

Longman
1724-1974

LONGMAN GROUP LIMITED London

*Associated companies, branches and representatives
throughout the world*

© Longman Group Limited 1974

All rights reserved. No part of this
publication may be reproduced, stored in
a retrieval system, or transmitted in any
form or by any means, electronic,
mechanical, photocopying, recording, or
otherwise, without the prior permission of
the Copyright owner.

First published 1974

ISBN 0 582 44134 X
Library of Congress
Catalog Card Number 73-85683

Printed in Great Britain by
William Clowes & Sons, Limited
London, Beccles and Colchester

AN ATLAS OF FUNGAL ULTRASTRUCTURE

A. Beckett
I. B. Heath &
D. J. McLaughlin

Longman

CONTENTS

Section 1a ('Flagellate lower Fungi')

Myxomycota & Mastigomycotina   1

Section 1b ('Non-Flagellate lower Fungi')

Zygomycotina   55

PREFACE

The aim of this atlas is to provide a collection of micrographs which illustrate the ultrastructural features of the major groups of fungi and which may be used to supplement information that is provided in existing mycology text books. The selection of material has been governed primarily by the availability of high-quality electron micrographs and secondly by the types of organisms frequently encountered in mycology courses and in mycology text books. As far as possible illustrations of various types of growth and reproduction have been included, but it must be emphasized that the examples are not comprehensive nor necessarily typical. In fact at present our knowledge of fungal ultrastructure is not sufficient to permit identification of the 'typical type'.

For *convenience* we have divided the atlas into three sections within which the fungal groups have been distributed as follows: Section 1. Lower Fungi: Myxomycota, Mastigomycotina and Zygomycotina; Section 2. Higher Fungi I: Ascomycotina and Deuteromycotina; Section 3. Higher Fungi II: Basidiomycotina. Each section is designed so that it may be read independently of the others although they are complementary and in many cases cross references have been made for the sake of clarity. Within each section the grouping of material on the basis of life-cycle stages has been selected in order to emphasize the similarities and variations of structure in response to specific problems. Such comparisons are not always apparent when complete life cycles of representative species are considered. Each section is provided with a brief introduction which summarizes and comments on some of the characteristic features of those fungi that are dealt with.

A glossary of selected ultrastructural terms is included at the end of the atlas for the benefit of those readers who are unfamiliar with the terminology. The collection and preparation of plates for this atlas was completed in December 1972.

A. BECKETT

I. B. HEATH

D. J. McLAUGHLIN

Acknowledgements

We are indebted to Dr M. G. Boyer, Dr R. Campbell, Professor L. E. Hawker, Dr Michele Heath, and Dr Esther McLaughlin for discussion and criticism of the manuscript and to D. S. Gunning, P. Barry-Calrow and J. McIntosh for their help in typing the manuscript.

We are especially grateful to the numerous workers who have contributed micrographs for this atlas and whose names appear with the relevant figure legends.

SECTION 1a

Myxomycota
& Mastigomycotina

Introduction

The 'Lower Fungi' are a collection of largely unrelated groups which are artificially associated in taxonomic treatments by virtue of the fact that they are either **unicellular** or possess an **aseptate vegetative mycelium**. Hyphal tip growth in mycelial forms is associated with the accumulation of **wall vesicles** at the apices of hyphae. An apical body or **Spitzenkörper** is not found.

Lower fungi may be sub-divided into flagellate and non-flagellate types.

Flagellate fungi as dealt with here include Myxomycota and Mastigomycotina, but it must be emphasized that the grouping of the Chytridiomycetes with the Hyphochytridiomycetes and Oömycetes within the sub-division Mastigomycotina is unreal and is not supported by recent work on their biochemistry, cell wall composition and ultrastructure. The use of biochemical and enzymic characters to identify related organisms and separate unrelated ones has shown that the Chytridiomycetes are more closely related to the Zygomycetes and higher fungi. In addition cell wall composition is different for the three classes. For example, the walls of Chytridiomycetes contain chitin and β glucan, those of Hyphochytridiomycetes contain cellulose and chitin while the walls of Oömycetes contain cellulose and β glucan. Although it is not the purpose of this atlas to present new taxonomic schemes it nevertheless deals with one body of evidence (ultrastructure) which may help in future reappraisals of the classification of lower fungi. Much of the information presented within this section relates to Oömycetes. This largely reflects the proportion of work done on the group and the availability of micrographs, but it should not be taken as necessarily representative of all lower fungi, as our present knowledge of all aspects of Oömycetes suggests that, with the exception of Hypochytridiomycetes, they are unrelated to other Mastigomycotina (and in many respects to other fungi) and should at least be given sub-divisional status equal in ranking to Zygomycotina, Ascomycotina and Basidiomycotina.

With this in mind we may consider some of the ultrastructural features of flagellate fungi.

As the term implies, these fungi possess **flagella** in their motile phases. A **centriole** also forms part of the **kinetosome** at the base of the flagellum in zoosporic and gametic stages, and in many genera centrioles are associated with the poles of an **intranuclear spindle** during nuclear division. In the motile stage of the life cycle many Oömycetes and Plasmodiophoromycetes possess an elaborate system of cisternae and vesicles which are concerned with osmoregulation of the cell. **Golgi dictyosomes** are part of this cell secretory apparatus. They have not been found in all lower fungi and are absent from higher fungi. Mitochondrial morphology in flagellate fungi, particularly Oömycetes, differs from that in Zygomycotina and higher fungi in that the cristae are tubular rather than flat plates and are usually orientated randomly within the lumen of the mitochondrion. Mitochondria of this type are also commonly found in many algal groups.

Non-flagellate lower fungi (Zygomycotina), in contrast, do not possess Golgi dictyosomes; the secretory processes in these fungi are performed, as in higher fungi, by individual Golgi cisternae. True centrioles have not been demonstrated in non-flagellate lower fungi. Related to the apparent absence of true centrioles and kinetosomes in Zygomycotina is the production, during asexual reproduction, of **endogenous, non-motile sporangiospores**. Studies on cell wall chemistry have shown a unique combination of **chitin** and **chitosan** in the cell walls of Zygomycotina.

For these reasons flagellate and non-flagellate fungi will be considered separately within this section.

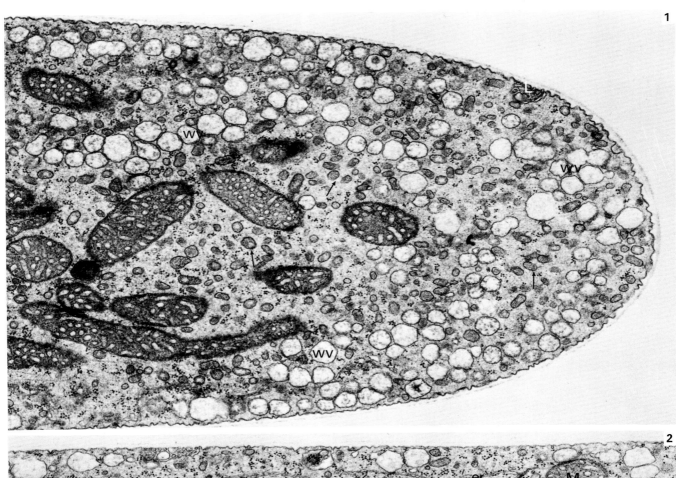

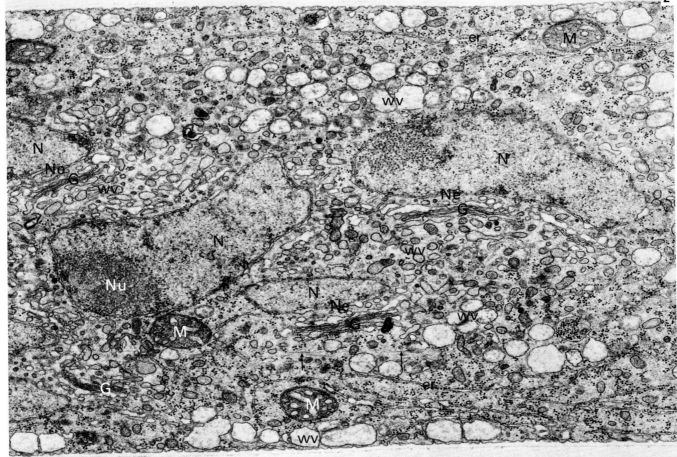

A. Vegetative Structures

Hyphae I. Cell wall synthesis

The vegetative colony of many Oömycetes and some Chytridiomycetes and Hyphochytridiomycetes is composed of tubular hyphae which extend by tip growth and multiply by sub-apical branching. A typical hyphal wall is composed of a random network of **fibrils**, usually cellulose or chitin (fig. 7) which are embedded in an **amorphous matrix** of glucans and proteins (see also Section II, p. 81, Section III, p. 187). Attempts to correlate ultrastructural observations of these components with the changes in cell wall plasticity necessary to account for tip growth are so far inconclusive.

Synthesis of the apical hyphal wall involves characteristic 'wall vesicles' which occur most abundantly in the apical 10 μm of a growing hypha (figs. 1, 4, 6) and also in localized sub-apical regions which are presumably sites of branch initiation (see Section II, p. 79; Section III, p. 185). Staining techniques (fig. 4) suggest that the wall vesicles contain **polysaccharide** (probably matrix material) which is released to the cell wall when the membranes of the vesicles fuse with the cell membrane. If this interpretation is correct, clearly the vesicle membranes contribute at least part of the expanding cell membrane needed to line the growing wall. Wall vesicles probably also contain the **enzymes** needed for synthesis of the cell wall and for softening a mature wall. The latter enzymes are expected at the sites of sub-apical branch initiation. Morphological and staining evidence suggests that wall vesicles are produced by the sub-apical **Golgi bodies** (figs. 2, 5, 6).

Membranous structures known as **lomasomes** (fig. 3) have frequently been implicated in cell wall synthesis. The most consistent feature to evolve from much controversy is the variability of both their structure and abundance. It has recently been suggested that they represent an excess of cell membrane relative to that needed to line the cell wall. Thus they would have no function beyond that attributable to the normal cell membrane. This hypothesis is not proven but fits the current observations better than other hypotheses.

Additional reading

GROVE, S. N. and BRACKER, C. E. (1970). Protoplasmic organization of hyphal tips among fungi: vesicles and Spitzenkörper. *J. Bact.*, **104**, 989.
HEATH, I. B., GAY, J. L. and GREENWOOD, A. D. (1971). Cell wall formation in the Saprolegniales: cytoplasmic vesicles underlying developing walls. *J. gen. Microbiol.*, **65**, 225.
HEATH, I. B. and GREENWOOD, A. D. (1970). The structure and formation of lomasomes. *J. gen. Microbiol.*, **62**, 129.

Fig. 1

Median longitudinal section of a hyphal tip of *Pythium aphanidermatum* (Edson) Fitzpatrick containing numerous wall vesicles (wv). The possible interrelationship between the large vesicles with lightly stained contents and the smaller vesicles (arrows) is unclear but both are probably involved in wall synthesis. Note a typical lomasome (L). Glutaraldehyde—osmium tetroxide fixation. × 26,000.

Fig. 2

Median longitudinal section of a sub-apical region of a hypha of *P. aphanidermatum*. Nuclei (N), with prominent nucleoli (Nu), and endoplasmic reticulum (er) are abundant in this region. The Golgi bodies (G), which in this species typically have their forming faces adjacent to the nuclear envelope (Ne), appear to be active in producing wall vesicles (wv). Microtubules (arrows) permeate the cytoplasm. Glutaraldehyde—osmium tetroxide fixation. × 27,000.

Figs. 1–2 From GROVE, S. N. and BRACKER, C. E. (1970). *J. Bact.*, **104**, 989.

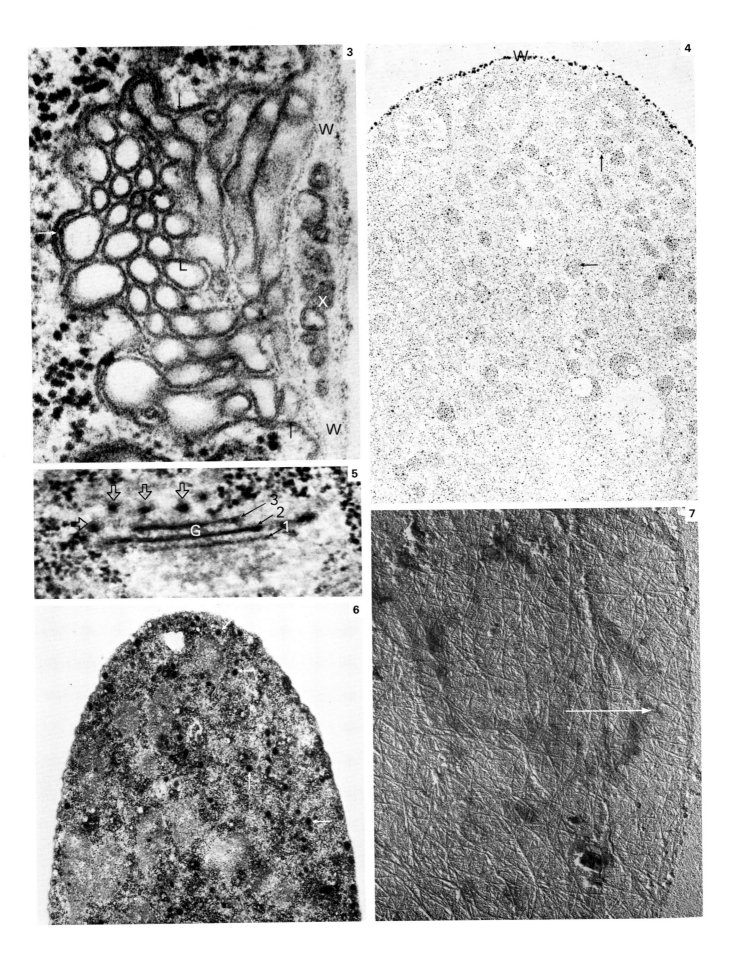

A. Vegetative Structures (cont.)

Hyphae I. Cell wall synthesis (cont.)

Fig. 3

A typical lomasome (L) in a hypha of *Saprolegnia ferax* (Gruithuisen) Thuret. Most of the polymorphic membranous tubules and vesicles lie in an invagination of the cell membrane (arrows) but some (X) appear to have become sequestered in the cell wall (W). Glutaraldehyde–osmium tetroxide fixation. × 112,700.

Fig. 4

Median longitudinal section of a hyphal tip of *S. ferax* treated with periodic acid and silver hexamine in order to demonstrate polysaccharide rich areas. The 'stain' (small black 'dots' of silver) is not specific but does occur more abundantly over the cell wall (W) and wall vesicles (arrows), suggesting that these areas have a higher concentration of polysaccharide than else-where in the section. Glutaraldehyde–osmium tetroxide fixation. × 39,400.

Fig. 5

Cross section of a Golgi body (G) in a hypha of *S. ferax* fixed only in glutaraldehyde then stained with uranyl acetate and lead citrate. There is an increasing amount of stain in successive cisternae (arrow), and the Golgi vesicles (open arrows) stain similarly to the wall vesicles seen in fig. 6. × 102,800.

Fig. 6

Median longitudinal section of a hyphal tip of *S. ferax* fixed as in fig. 5. The wall vesicles (arrows) stain very intensely after this unusual treatment. × 17,800.

Fig. 7

Portion of the apex (in the direction of the arrow) of a *S. ferax* hyphal wall from which the amorphous material has been re-moved, thus exposing the random network of cellulose fibrils which occur all over the hypha and completely enclose the apex. Pd/Au shadowed. × 41,400.

Figs. 3–7   *3 From* HEATH, I. B. and GREENWOOD, A. D. (1970). *J. gen. Microbiol.,* **62**, 129. *4–6 From* HEATH, I. B., GAY, J. L. and GREENWOOD, A. D. (1971). *J. gen. Microbiol.,* **65**, 225. *7 Micrograph by* I. B. HEATH, York University, Ontario.

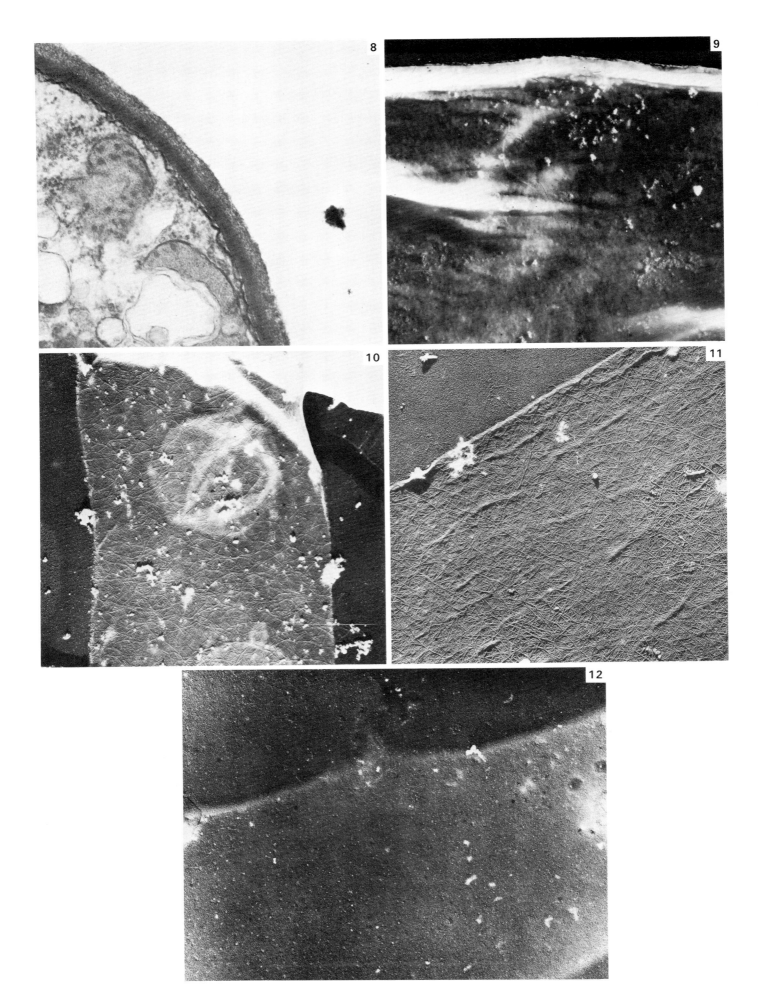

A. Vegetative Structures (cont.)

Hyphae II. Cell wall structure

Enzymic-dissection has been used to investigate hyphal wall structure in some fungi for which chemical analyses of the hyphal walls are available. Thus major constituents of the walls of the Oömycete *Phytophthora parasitica* (Persoon) de Bary are known to be **cellulose**, **protein** and a **glucan** containing $\beta1,3$ and $\beta1,6$ linkages. The technique consists of applying enzymes, singly and then sequentially, to living mycelial material. After the enzymic treatments, the material is examined either in shadow cast preparations or in section. The former enables one to detect any changes in surface topography and the latter enables one to measure any changes in wall thickness produced as a result of the enzymic treatments. This method has been used on *P. parasitica*. The enzymes used were **cellulase**, **pronase**, and **laminarinase** (a $\beta1,3:\beta1,6$ glucanase). Untreated walls (fig. 8) show a two-layered structure in section with a more electron dense inner layer. Shadowed preparations of walls which have been incubated only in buffer have an amorphous appearance (fig. 9); such walls given single treatments with cellulase or pronase have a similar appearance. Laminarinase treatment, however, removes the outer wall region to reveal **microfibrils** of cellulosic

dimensions (fig. 10). Only the outermost, predominantly transverse, microfibrils are revealed by this treatment, the underlying ones are obscured by amorphous material. If laminarinase treatment is followed by pronase treatment, then not only do the surface microfibrils become more sharply defined, but underlying randomly arranged microfibrils are revealed as well (fig. 11). This is thought to be due to removal of an enmeshing layer of protein. If, however, laminarinase treatment is followed by cellulase treatment, then the microfibrils are removed and a hyphal 'ghost' is left (fig. 12) which is presumably proteinaceous in nature. Both the sequences laminarinase/cellulase/pronase and laminarinase/pronase/cellulase leave no identifiable remnants in shadow cast preparations and, in section, only the plasmalemma may be seen. Figure 13 summarizes the information obtained on hyphal wall structure of *P. parasitica* and compares this model of wall structure with that seen in sections of untreated walls. The reconstruction embodies information obtained from electron micrographs of wall sections after enzymic treatments which are not illustrated here.

Additional reading

BARTNICKI-GARCIA, S. (1966). Chemistry of hyphal walls of *Phytophthora*. *J. gen. Microbiol.*, **42**, 57.
HUNSLEY, D. and BURNETT, J. H. (1970). The ultrastructural architecture of the walls of some hyphal fungi. *J. gen. Microbiol.*, **62**, 203.

Fig. 8

Cross section through part of the wall of an untreated cell of *Phytophthora parasitica* showing two layers, the inner one being more electron dense than the outer one. Glutaraldehyde–osmium tetroxide fixation. × 40,000.

Fig. 9

A shadowed preparation of cell walls of *P. parasitica* which have been incubated only in buffer. Note the amorphous appearance. Pd/Au shadowed 40/60 Cot[1–3]. × 15,000.

Fig. 10

A shadowed preparation of cell walls of *P. parasitica* which have been treated with laminarinase. Note presence of microfibrils. Pd/Au shadowed 40/60 Cot[1–3]. × 15,000.

Fig. 11

A shadowed preparation of cell walls of *P. parasitica* which have been treated with laminarinase followed by pronase. Randomly arranged microfibrils may be seen underlying the sharply defined surface microfibrils. Pd/Au shadowed 40/60 Cot[1–3]. × 15,000.

Fig. 12

A shadowed preparation of cell walls of *P. parasitica* which have been treated with laminarinase followed by cellulase. All microfibrils have been removed, leaving a hyphal 'ghost', which is presumably proteinaceous. Pd/Au shadowed 40/60 Cot[1–3]. × 15,000.

Figs. 8–12 *8, 11 Micrographs by* DR D. HUNSLEY, Department of Agricultural Science, University of Oxford. *9, 10, 12 From* HUNSLEY, D. and BURNETT, J. H. (1970). *J. gen. Microbiol.*, **62**, 203.

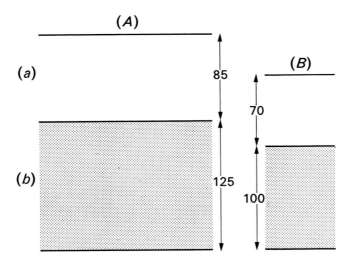

A. Vegetative Structures (cont.)

Hyphae II. Cell wall structure (cont.)

Fig. 13

(A) Reconstruction of section through the wall of a hypha from a
5-day culture, based on enzymic-dissection experiments. The
numbers represent the mean thickness of the layers in nm.
(*a*) Outermost layer of amorphous glucan containing β1,3 and
β1,6 linkages. (*b*) Innermost layer of cellulose microfibrils and
protein.

(B) Wall layers visible in a section of an untreated wall from a
5-day culture, fixed in glutaraldehyde–osmium tetroxide. The
inner layer is more electron opaque. The numbers represent the
mean thickness of the layers in nm.

Fig. 13 *From* HUNSLEY, D. and BURNETT, J. H. (1970). *J. gen.
Microbiol.*, **62**, 203.

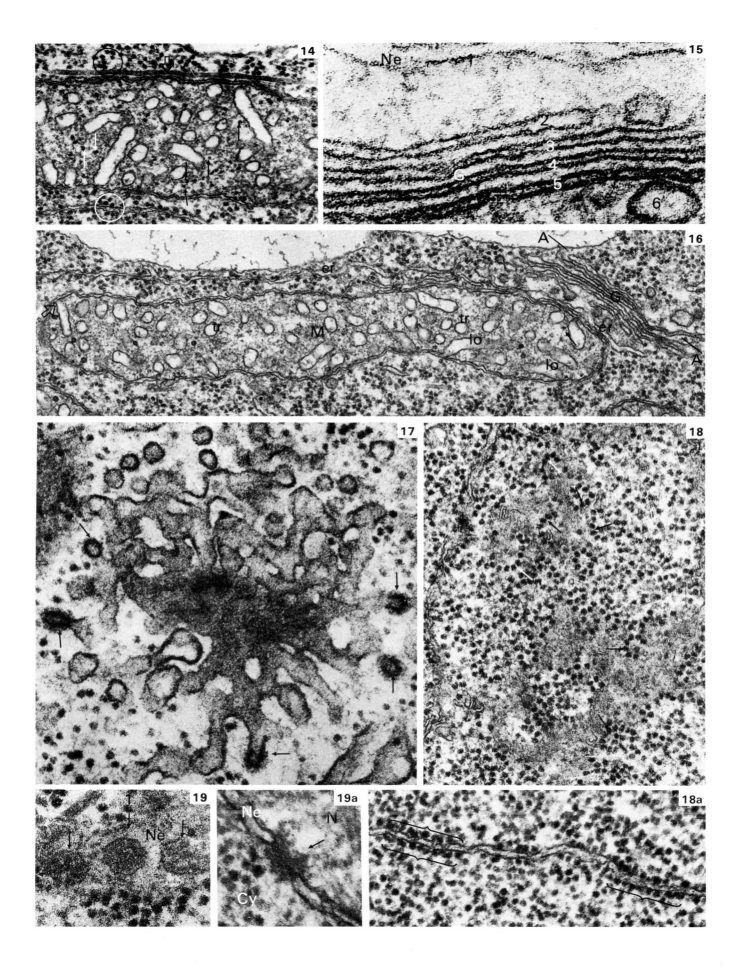

A. Vegetative Structures (cont.)

Hyphae III. Cell components

Mitochondria occur throughout the vegetative cytoplasm of Oömycetes with the exception of the apical 10 μm but they are most abundant immediately below this region. Their cristae are tubular (figs. 14, 16) and their matrices contain both ribosome-like particles (fig. 14) and fibrils which are comparable to the DNA fibrils of other organisms (fig. 16). Each mitochondrion is about 1 μm in diameter but, whilst those near the apex are only 1–2 μm long, those in the older regions of the hyphae may be 15 μm or more in length, a difference which suggests formation (by division?) near the apex and subsequent elongation.

Nuclei occur below the main mitochondrial zone. Each interphase nucleus (fig. 2) contains a nucleolus and the nuclear envelopes bear numerous **nuclear pores** (figs. 19, 19a) which are characteristically plugged with osmiophilic material. Continuous with the nuclear envelope is an extensive network of fenestrated sheets of double membranes known as the **endoplasmic reticulum** (figs. 18, 18a). Most of the endoplasmic reticulum has ribosomes attached to the cytoplasmic side of the membranes and thus may be termed rough endoplasmic reticulum.

Golgi bodies or **dictyosomes** are present in all Oömycetes so far examined. They are typically composed of 4 or 5 flattened cisternae with their forming faces adjacent to either the nuclear envelope or a cisternum of the endoplasmic reticulum (figs. 2, 16, and 17). The change in staining properties of the lumina of the cisternae (fig. 5) and the progressive change in membrane structure across the cisternal stacks (fig. 15) supports the hypothesis that lumen and membranes are formed in the endoplasmic reticulum, are modified as they pass through the Golgi bodies, and are finally packaged for transport to the cell surface in a Golgi vesicle whose membrane is capable of fusion with the cell membrane. The necessity for membrane transformation in this flow system is seen when it is considered that the membrane of the endoplasmic reticulum differs considerably in structure from that of a cell membrane.

Microtubules permeate hyphal cytoplasm, running in relatively straight lines for many micrometers parallel to the long axis of the hypha (fig. 2). These 25 nm diameter tubules are commonly associated with nuclei (fig. 45) and mitochondria (fig. 14) in a manner which has led to the suggestion that an interaction between the organelle and the microtubule plays a role in the commonly observed phenomenon of organelle motility which is independent of cytoplasmic streaming (see also p. 23 and Section II, p. 75).

In addition to the above components, hyphae in most lower fungi often contain **ribosomes** (figs. 18, 18a), **microbodies** (fig. 22) and **lipid droplets** (fig. 33), the latter being the major food storage material in these fungi.

Additional reading

GROVE, S. N., BRACKER, C. E. and MORRÉ, D. J. (1968). Cytomembrane differentiation in the endoplasmic reticulum – Golgi apparatus – vesicle complex. *Science, N.Y.*, **161**, 171.

Fig. 14

Mitochondrion of *Thraustotheca clavata* (de Bary) Humphrey showing numerous mitochondrial ribosomes (arrows), which appear smaller than those of the cytoplasm (circled). Note the associated microtubule (open arrow). × 111,200.

Fig. 15

Cross section of a Golgi body of *Pythium ultimum* Trow showing the way in which the structure of the membranes changes from that of the nuclear envelope (which is comparable to the endoplasmic reticulum) (Ne, 1), through the cisternae of the Golgi body (G, 2, 3, 4 and 5), and finally to that of the Golgi vesicle (6). Glutaraldehyde–osmium tetroxide fixation. × 225,000.

Fig. 16

Mitochondrion (M) of *Saprolegnia ferax* (Gruithuisen) Thuret showing cross sectioned (tr) and longitudinally sectioned (lo) tubular cristae (open arrow shows where a crista is clearly an invagination of the inner membrane), and DNA-like fibrils (solid arrow). A Golgi body (G) which, as is typical for this species, is associated with the endoplasmic reticulum (er) and the mitochondrion. × 50,200.

Fig. 17

Surface section of a Golgi cisternum of *S. ferax* (i.e. sectioned in a plane similar to that marked by the lines A–A in fig. 16). The cisternum is typically fenestrated at the margin and has characteristic osmiophilic vesicles attached to it (arrows). Osmium tetroxide fixation. × 102,600.

Fig. 18

Surface section of a cisternum of the endoplasmic reticulum of *T. clavata*. The clear, uniformly 'grey' areas are where the section passed through the membrane or lumen of the cisternum. Curved 'chains' of polyribosomes attached to the membrane are indicated by arrows. × 74,000.

Fig. 18(a)

Cross sectioned cisternum of endoplasmic reticulum of *T. clavata* with several polyribosome complexes (between brackets). × 92,000.

Fig. 19

Surface section of the nuclear envelope of *T. clavata* showing three nuclear pores (arrows), one of which (No. 1) is sectioned medially. × 166,500.

Fig. 19(a)

Cross section of a nuclear pore (arrow) in the nuclear envelope (Ne) of *T. clavata*. Note the osmiophilic 'plug' in the pore between the cytoplasm (Cy) and the nucleoplasm (N). × 137,400.

Figs. 14–19(a) *(14, 16, 18, 18(a), 19, 19(a) Glutaraldehyde–osmium tetroxide fixation.) Micrographs by* I. B. HEATH, York University, Ontario. *15 From* GROVE, S. N., BRACKER, C. E. and MORRÉ, D. J. (1968). *Science, N.Y.*, **161**, 171. Copyright 1968 by the American Association for the Advancement of Science. *17 From* HEATH, I. B. and GREENWOOD, A. D. (1971). 'Ultrastructural observation on the kinetosomes, and golgi bodies during the asexual life cycle of saprogegnia', *Z. Zellforsch. mikrosk. Anat.*, **112**, 371–89. Berlin–Heidelberg–New York: Springer.

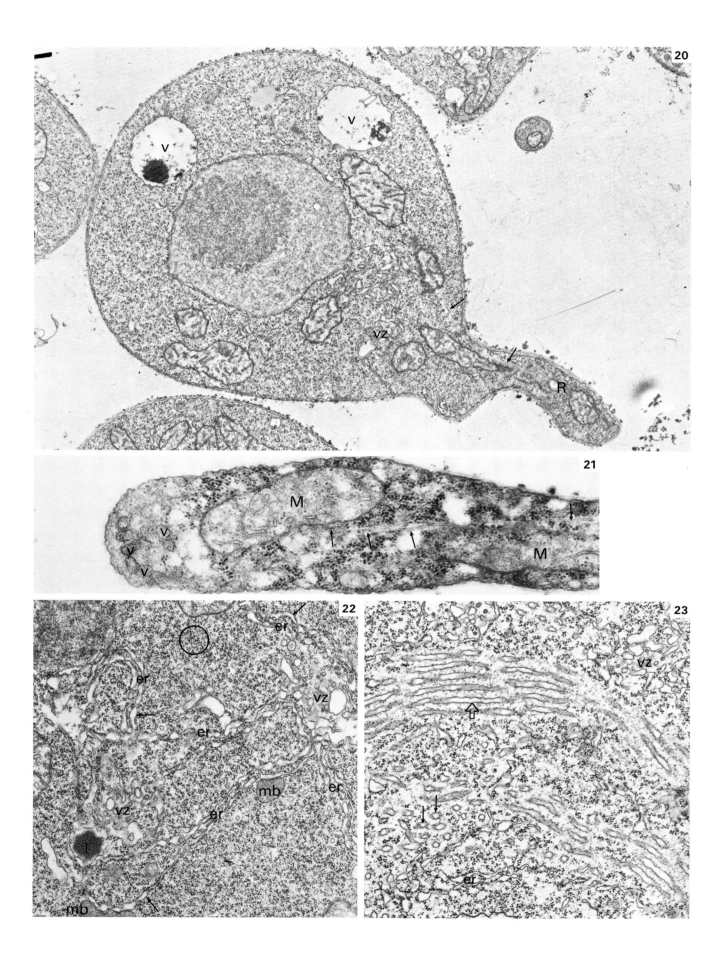

A. Vegetative Structures (cont.)

Monocentric type thallus

In many Chytridiomycetes, e.g. *Blastocladiella emersonii* Cantino and Hyatt, the **monocentric** thallus is characterized by a large spherical, multinucleate 'cell' attached to the substrate by a small, branching anucleate network of **rhizoids** (fig. 20) which contrasts with the hyphal type thallus typical of most Oömycetes (see p. 5). From the juvenile stage shown in fig. 20 the spherical part of the thallus expands considerably. Clearly such expansion necessitates extensive cell wall synthesis, but to date 'wall vesicles' as found in the hyphal tip system (see p. 5) are lacking, suggesting that a different synthetic process is operating. Unlike the body of the thallus which grows by non-localized wall expansion, **rhizoids** extend by tip growth and do contain vesicles at their apices (fig. 21), but at present we only have circumstantial evidence that these vesicles are comparable in function with those in hyphal apices (see p. 5).

Expansion of the body of the thallus is accompanied by mitosis, proliferation of mitochondria and cytoplasm, and the formation of an extensive membrane system (fig. 22). Morphologically typical Golgi bodies are lacking but associated with the endoplasmic reticulum are characteristic **'vesicular zones of exclusion'** (fig. 22) which are thought to be functional equivalents of the Golgi bodies. Towards the end of the growth phase of the plant, prior to zoosporulation, highly characteristic bundles of membranous tubules become apparent in the cytoplasm (fig. 23). At present we have little indication of their function. They are included in this introduction as an illustration of the surprises and new problems which confront investigators of fungal ultrastructure.

Additional reading

LESSIE, P. E. and LOVETT, J. S. (1968). Ultrastructural changes during sporangium formation and zoospore differentiation in *Blastocladiella emersonii*. Am. J. Bot., **55**, 220.

Fig. 20

A young thallus of *Blastocladiella emersonii* Cantino and Hyatt, shown developing from an encysted zoospore. At this early stage the mature form of spherical body and basal rhizoids (R) is apparent. The vesicles (v) are remnant structures from the zoospore, their function is unknown but their structure suggests that they should be compared with the dense bodies described on p. 31. Note the microtubules (arrows) which appear to be 'guiding' the mitochondria down the rhizoid. A 'vesicular zone of exclusion' (vz) is already apparent. × 15,800.

Fig. 21

Longitudinal section of the tip of a growing rhizoid of *B. emersonii* showing the apical cluster of vesicles (v) and also mitochondria (M) which appear to be associated with microtubules (arrows) (cf. fig. 14). × 37,750.

Fig. 22

Portion of the body of an older thallus than that shown in fig. 20. Extensive 'sheets' of endoplasmic reticulum (er), some dilated (arrows), have developed and in some cases these are associated with microbodies (mb) and lipid droplets (l). Ribosomes (e.g. circled area) are abundant in the cytoplasm but are excluded from the 'vesicular zones of exclusion' (vz). × 14,600.

Fig. 23

Portion of an old thallus of *B. emersonii* immediately prior to zoosporulation. Characteristic membranous tubules are seen in both longitudinal (open arrows) and transverse (arrows) section. Endoplasmic reticulum (er) and a 'vesicular zone of exclusion' (vz) are present. × 23,800.

Figs. 20–3 *Glutaraldehyde–osmium tetroxide fixation. Micrographs by* DRS W. E. BARSTOW *and* J. S. LOVETT, *Purdue University, Lafayette, Indiana.*

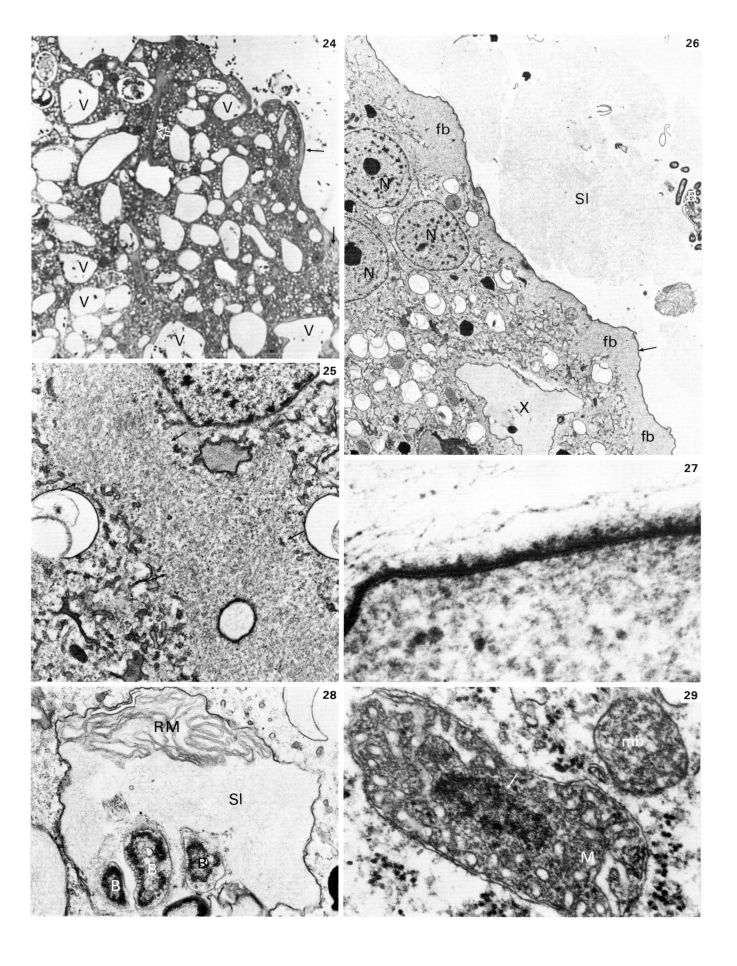

A. Vegetative Structures (cont.)

Myxomycete plasmodia

The major point of difference between the structure of a vegetative hypha and a Myxomycete plasmodium is the absence of a rigid cell wall. The plasmodium is typically bounded by a thickened cell membrane (fig. 27) and an extracellular layer of fibrous **slime** (figs. 26, 27) which is secreted by the plasmodium. The precise role of the slime is uncertain. It may act as a protective lubricant as the plasmodium streams over its substrate, or, since it is highly elastic, it may interact in some way with the cytoplasm to help produce the vigorous cytoplasmic streaming so characteristic of these organisms.

Myxomycete plasmodia contain a **fibrous protein** comparable to muscle actin. Hence there have been numerous ultrastructural attempts to locate this substance in the cytoplasm and to determine if its arrangement can help to explain the characteristic pulsating streaming found in many plasmodia. The **cortical cytoplasm** immediately adjacent to the cell membrane is usually devoid of organelles (figs. 24, 26) and contains abundant fibrils with a diameter of approximately 5–8 nm. These fibrils may be arranged both longitudinally and circumferentially relative to the long axis of the plasmodium, when such an axis exists. They are thought to be the actin-like protein and it has been suggested that their appropriate contraction could produce the observed **streaming** of the contained cytoplasm. There are also bundles of similar fibrils which run through the non-peripheral cytoplasm (figs. 24, 25). Their role is more difficult to explain at present. The main body of cytoplasm of the plasmodium contains abundant **nuclei** (fig. 26), **mitochondria** with characteristic **DNA cores** (fig. 29), and **food vacuoles** in which the food bacteria, ingested by pinocytosis, are digested (fig. 28).

Additional reading

RHEA, R. P. (1966). Electron microscopic observations on the slime mold *Physarum polycephalum* with specific reference to fibrillar structures. *J. Ultrastruct. Res.*, **15**, 349.

DANIEL, J. W. and JARLFORS, U. (1972). Plasmodial ultrastructure of the myxomycete, *Physarum polycephalum*. *Tissue & Cell*, **4**, 15.

Fig. 24

Light micrograph of a portion of a plasmodium of *Physarella oblonga* (Berk. and Curt.) Morgan showing a fibre (open arrow) and the fibrous material at the periphery of the plasmodium (arrows). Food vacuoles (V) containing bacteria (small black dots) are abundant. × 500.

Fig. 25

Electron micrograph of a part of the fibre shown in fig. 24. Organelles are excluded from this fibril-containing region (between arrows). × 15,300.

Fig. 26

Part of the edge of a plasmodium of *P. oblonga* showing the peripheral fibrous layer of material (fb) beneath the cell membrane (arrow). Bacteria (B) are trapped in the extracellular slime (Sl). The structure marked with an X is probably a food vacuole but only serial sectioning would prove that it was not an invagination of the cell membrane and thus extracellular space. Three nuclei (N) are also present. × 5,100.

Fig. 27

Detail of the cell membrane of *P. oblonga* showing the osmiophilic elaboration of the outer part of the membrane. Compare this cell membrane with that of fig. 3. × 126,000.

Fig. 28

Food vacuole of *P. oblonga* containing as yet undigested bacteria (B), extracellular slime (Sl) which was apparently trapped during pinocytosis and residual membranous material (RM) which may represent indigestible membranes of previously digested bacteria. × 19,200.

Fig. 29

A typical myxomycete mitochondrion (M) containing an osmiophilic core of DNA (arrow) (cf. fig. 16) and branched tubular cristae. 'mb' is a microbody. × 90,000.

Figs. 24–9 *Glutaraldehyde–osmium tetroxide fixation. Micrographs by* DR H. C. ALDRICH, *University of Florida, Gainesville, Florida.*

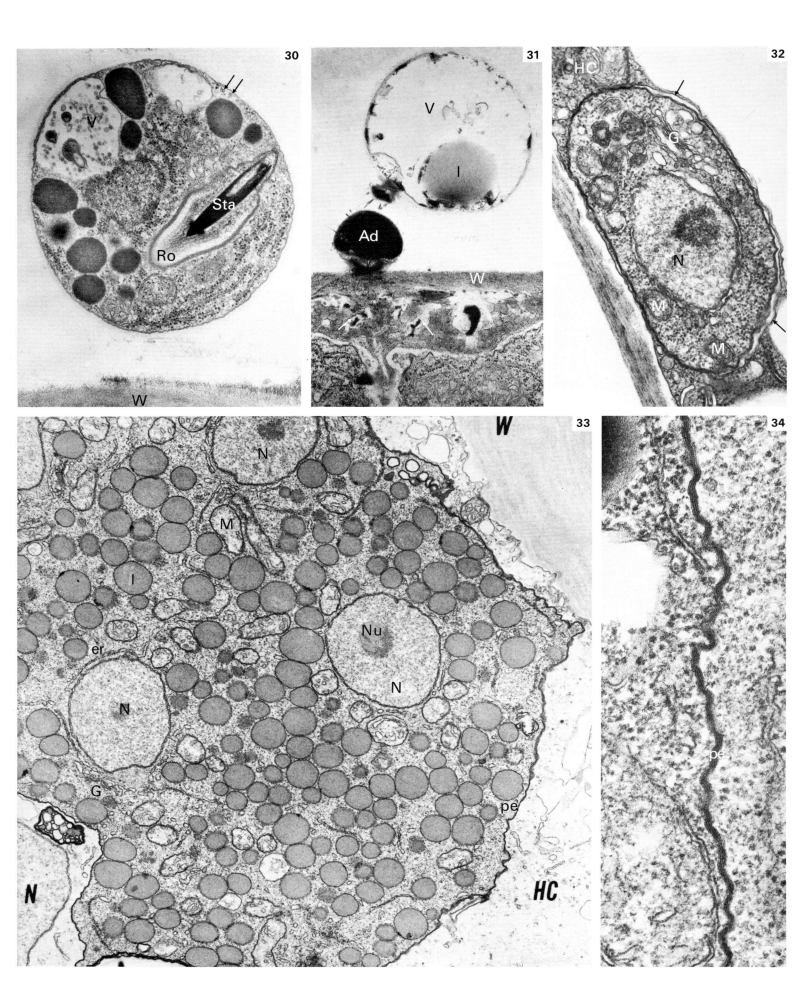

A. Vegetative Structures (cont.)

Plasmodiophoromycete plasmodia

Plasmodiophora brassicae Woron. represents a highly specialized group of obligate parasites of higher plants, the Plasmodiophoromycetes. In this group the vegetative thallus is an intracellular multi-nucleate **plasmodium** which is separated from the surrounding host cytoplasm by two closely appressed membranes, presumably one of fungus and one of host origin (figs. 33, 34). This host—parasite interface should be compared with the functionally comparable haustorium discussed on p. 21. Both are involved, at least in part, in nutrient transfer.

In order to infect a new host the parent plasmodium cleaves into uninucleate zoospores which are released from the old host, swim through the water film in the soil, and encyst on a new root hair. The mode of penetration is a wonderful example of adaptation to a specific problem. A bullet shaped **'Stachel'** develops within the cyst (fig. 30). When the 'Stachel' is mature the cyst produces a tubular extension, the **'adhesorium'** (fig. 31) from which the 'Stachel' is 'fired' through the host wall into its cytoplasm. Within 12 seconds the fungal cytoplasm enters through the hole made by the 'Stachel', is swept away by the streaming host cytoplasm (fig. 32) and subsequently develops into another plasmodium. Within 3 minutes after penetration of the 'Stachel' the host produces a localized deposition of material containing **callose** around the site of penetration. This is a common wound reaction, but in this case it is clearly too late to prevent infection (see also p. 21).

Additional reading

AIST, J. R. and WILLIAMS, P. H. (1971). The cytology and kinetics of cabbage root hair penetration by *Plasmodiophora brassicae*. Can. J. Bot., **49**, 2023.
WILLIAMS, P. H. and MCNABOLA, S. S. (1967). Fine structure of *Plasmodiophora brassicae* in sporogenesis. Can. J. Bot., **45**, 1665.

Fig. 30

Encysted spore of *P. brassicae* on the host cell wall (W), showing a nearly mature 'Stachel' (Sta) lying in an invagination of the cell membrane, called the 'Rohr' (Ro). Doublet microtubules (arrows) remain from the withdrawn flagellum axoneme. × 36,650.

Fig. 31

Remains of a *P. brassicae* cyst whose cytoplasm, excluding a lipid droplet (I), has been injected through the host wall (W). The 'adhesorium' (Ad) is formed by eversion of the 'Rohr'. The pressure needed to evert the 'adhesorium' and possibly to fire the 'Stachel' and extrude the amoeba into the host is probably produced by expansion of the cyst vacuole (V, fig. 30), which then remains in the cyst. After the cyst cytoplasm enters the host a mound of reaction material (arrows) is produced around the breach in the host wall. × 23,500.

Fig. 32

A uninucleate amoeba of *P. brassicae* shortly after entry into the host cytoplasm (HC). A nucleus (N), Golgi body (G) and mitochondria (M) are present. The amoeba is surrounded by a seven layered membrane complex (cf. fig. 34) and additionally in part by a flattened vesicle (arrow) of unknown function. × 32,250.

Fig. 33

A multinucleate (N) plasmodium of *P. brassicae* in a host cell (HC). The plasmodium contains numerous lipid droplets (I), mitochondria (M), endoplasmic reticulum (er) and Golgi bodies (G), all enclosed by a seven-layered membrane, the plasmodial envelope (pe). Glutaraldehyde—osmium tetroxide fixation. × 15,800.

Fig. 34

A detail of the plasmodial envelope (pe) of *P. brassicae* showing the seven alternating dark and light layers, i.e. two, three layered ('dark-light-dark') membranes separated by a 'light' space. Glutaraldehyde—osmium tetroxide fixation. × 75,000.

Figs. 30–4 *(30–2 Glutaraldehyde—osmium tetroxide fixation) From* AIST, J. R. and WILLIAMS, P. H. (1971). Can. J. Bot., **49**, 2023. Reproduced by permission of the National Research Council of Canada. *33 From* WILLIAMS, P. H. and MCNABOLA, S. S. (1967). Can. J. Bot., **45**, 1665. *34 From* WILLIAMS, P. H. and MCNABOLA, S. S. (1970). Phytopathology, **60**, 1557.

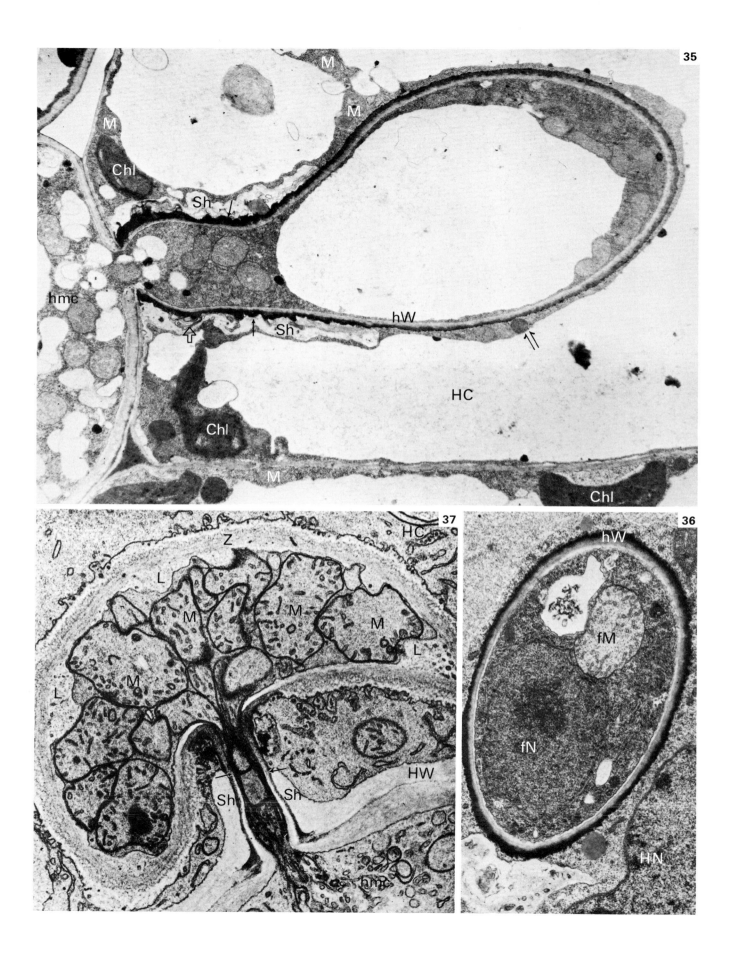

A. Vegetative Structures (cont.)

Haustoria

Many of the hyphal flagellate fungi are plant parasites. Of these the most highly specialized are the obligately parasitic Peronosporaceae which have haustoria. A haustorium is a specialized hyphal branch which presumably enables the fungus to extract nutrients from a host cell without causing host death before the fungus has reproduced. Ultrastructural details of haustoria vary from species to species, for example, *Peronospora parasitica* (Fr.) Tul. has a haustorium which is completely enclosed by an extension of the **haustorial mother cell** wall (fig. 35), whereas in *Albugo candida* (Pers.) Kuntze this wall only reaches a short way up the neck of the haustorium (fig. 37). Instead of this wall the haustorium of *A. candida* is surrounded by a **'zone of apposition'**,[1] material of uncertain nature and origin, probably from both host and fungus. Similarly the contents of haustoria may vary. Haustoria of *P. parasitica* contain all the organelles typical of the hyphae (figs. 35, 36), whilst those of *A. candida* lack nuclei but are rich in mitochondria and lomasomes (fig. 37). At present few of these observations can be correlated with functions related to the specialized role of the haustorium.

No discussion of haustoria would be complete without considering the host reaction. The typical response is the rapid synthesis of some form of **sheath** or **collar** (figs. 35, 37). In *A. candida* this sheath is continuous with the cell wall (fig. 37) but of unknown composition, but in *P. parasitica* (fig. 35) it is rich in **callose**, a polysaccharide characteristically produced in response to wounding in plants. If the sheath is a defence mechanism it is unsuccessful in most infections.

[1] Structures associated with haustoria have been variously named in the literature by different workers. The **'zone of apposition'** (Peyton and Bowen 1963) may be equated to the **'encapsulation'** (Berlin and Bowen 1964) and the **'haustorial sheath'** (Bracker 1968, Littlefield and Bracker 1972; see additional reading Section II, p. 91; Section III, p. 197). The **'sheath'** (Peyton and Bowen 1963, Berlin and Bowen 1964) may be equated to the **'collar'** (Bracker 1968; see additional reading Section II, p. 91).

Additional reading

BERLIN, J. D. and BOWEN, C. C. (1964). The host parasite interface of *Albugo candida* on *Raphanus sativus*. *Am. J. Bot.*, **51**, 445.

CHOU, C. K. (1970). An electron microscope study of host penetration and early stages of haustorium formation of *Peronospora parasitica* (Fr.) Tul. on cabbage cotyledons. *Ann. Bot.*, **34**, 189.

PEYTON, G. A. and BOWEN, C. C. (1963). The host–parasite interface of *Peronospora manshurica* on *Glycine max*. *Am. J. Bot.*, **50**, 787.

Fig. 35

A longitudinal section of a haustorium of *Peronospora parasitica* (Fr.) Tul. in a cabbage cell (HC). The haustorium wall (hW) appears to be an extension of the haustorial mother cell (hmc) wall. Between the haustorial wall and the host cell membrane is a layer of osmiophilic material which is most prominent within the sheath (Sh) (arrows). This material, probably phospholipid, is associated with proliferations of membrane (open arrow), some of which appear to be trapped in the sheath material. The large vacuole in the haustorium is characteristic of an old haustorium, yet the invaded cell (HC) at this stage does not appear to differ significantly from adjacent uninfected cells; chloroplasts (Chl), mitochondria (M) and vacuolar membrane (double arrow) are all intact. × 10,500.

Fig. 36

Portion of a cross sectioned haustorium of *P. parasitica* showing a fungal nucleus (fN) and mitochondrion (fM) and also a host nucleus (HN) which is typically located adjacent to the haustorium. × 18,500.

Fig. 37

A typical haustorium of *Albugo candida* (Pers.) Kuntze in a radish cell (HC). Compare the characteristic size and shape with fig. 35. The sheath (Sh) is an extension of the host cell wall (HW). The wall of the haustorial mother cell (hmc) terminates halfway up the neck of the haustorium (arrows). The body of the haustorium is bounded by a 'zone of apposition' (Z) of uncertain origin. Lomasomes (L) and mitochondria (M) are typically abundant. Unlike most illustrations in this section, this material was fixed in potassium permanganate which only preserves cell walls and membranes. Hence these structures are more prominent than in the material fixed in glutaraldehyde and osmium tetroxide (e.g. figs. 35, 36). × 20,300.

Figs. 35–7 *(35, 36 Glutaraldehyde–osmium tetroxide fixation)* *Micrographs by* DR J. SARGENT, Unit of Developmental Botany, A.R.C., Cambridge University. *37 From* BERLIN, J. D. and BOWEN, C. C. (1964). *Am. J. Bot.*, **51**, 445.

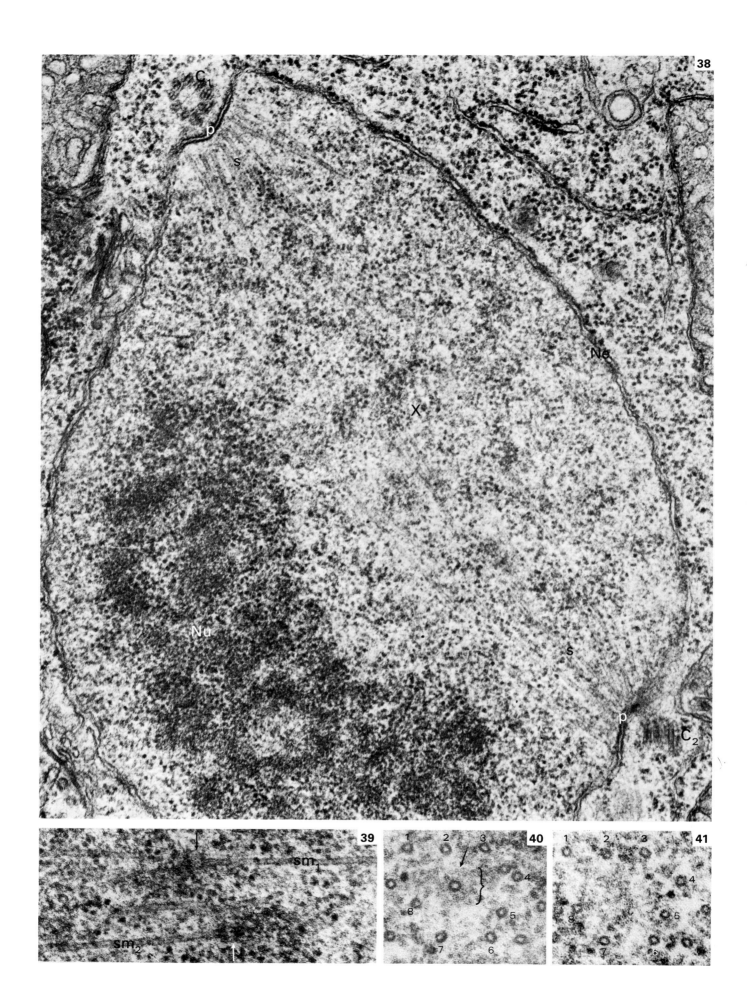

A. Vegetative Structures (cont.)

Mitosis

The mechanism of mitosis in many flagellate fungi has been the subject of considerable controversy for many years. Three examples will be described here in order to point out the range of variation so far described.

In the Oömycete, *Saprolegnia ferax* (Gruithuisen) Thuret each interphase nucleus is accompanied by a pair of **centrioles** which are aligned end to end in contrast to the 90° configuration typical of most organisms. Centriole replication (fig. 42) precedes mitosis, subsequently the two centriole pairs move apart as a small **microtubular spindle** develops between 'pockets' of the nuclear envelope (fig. 43). As centriole migration continues around the nucleus so the spindle elongates (fig. 38). **Kinetochores** (centromeres) are at first found at the equator of the spindle (figs. 39, 40, 41) but as the spindle continues to elongate (fig. 46) they can be detected at the poles of the spindle. (Because the chromosomes do not stain clearly in this species kinetochores can only be defined as characteristic microtubule terminations whose behaviour during mitosis is the same as that of kinetochores [see glossary].) As the spindle continues to elongate so the nucleus becomes hour-glass shaped (fig. 46). Division is completed by constriction of the narrow isthmus, presumably by fusion of the nuclear envelope. The nuclear envelope and **nucleolus** persist morphologically intact throughout division, the latter apparently dividing in a similar manner to the nucleoplasm. A novel association between the **nuclear envelope** and some of the **cytoplasmic microtubules** (figs. 44, 45) is thought to play a role in providing the forces needed to elongate the nucleus during division and also in moving the nuclei through the hyphae during interphase.

Mitosis in the Chytridiomycete *Catenaria anguillulae* Sorokin is more comparable to that of higher plants and animals in that the chromosomes stain well and can be seen to form a well-defined **metaphase plate** (fig. 48). However, the nuclear envelope remains intact during most of the mitotic process (unlike higher organisms where it disperses at prophase), the spindle microtubules diverging from polar 'pocket' like regions of the nuclear envelope as in *S. ferax* (figs. 48, 49). Adjacent to each 'pocket' is a pair of centrioles (fig. 47) which have the more normal 90° orientation (contrast with *S. ferax*). Whilst kinetochores

have not been demonstrated, anaphase separation of the chromosomes is known to involve elongation of the pole to pole spindle microtubules (**continuous microtubules**) in the absence of concomitant shortening of the pole to chromosome tubules (**chromosomal microtubules**) (figs. 49, 50). Final separation of daughter nuclei is accomplished by an extremely unusual mechanism whereby the chromosomes round up into two small masses at each end of the nucleus. Then the nuclear envelope constricts around these masses, excluding the main part of the nucleoplasm (fig. 51) which soon breaks down. In contrast to *S. ferax*, the **nucleolus** is lost at the onset of mitosis and there is no evidence for a nuclear envelope–microtubule association. The fact that the spindle is larger in *C. anguillulae* may mean that this additional force producing mechanism is unnecessary.

In the Myxomycete, *Physarum flavicomum* Berk., mitosis exhibits the unusual condition of differing markedly in the **haploid** and **diploid** phases of the life cycle. In the diploid plasmodium **centrioles** are absent from the poles of the spindle (fig. 53); during prophase the nucleolus disperses and then a **metaphase plate** develops on a well formed microtubular spindle within the intact nuclear envelope (fig. 53). During chromosome migration at anaphase the nuclear envelope remains intact at the equator of the nucleus but begins to break down at the poles where some spindle microtubules pass out into the cytoplasm (fig. 54). At telophase the **chromosomes** group together at each pole and the nuclear envelope begins to reform around each group, thus excluding the spindle microtubules which break down in the cytoplasm (fig. 55). At present no details of possible kinetochores are available.

In contrast to this process, mitosis in the haploid myxamoebae (see p. 49) is accomplished with well defined **centriole pairs** at each spindle pole (fig. 52) and a nuclear envelope and nucleolus which disperse at prophase. Only when cytokinesis (myxamoebae are uninucleate in contrast to the coenocytic plasmodium) is nearly complete does the nuclear envelope reform around each group of chromosomes. Details of chromosome separation appear to be similar to the process in the plasmodia and in *C. anguillulae*.

Fig. 38

A vegetative nucleus of *Saprolegnia ferax* at a stage comparable to metaphase. The nuclear envelope (Ne) encloses the persistent nucleolus (Nu) and a small (in numbers of tubules) microtubular spindle (s) which converges to polar 'pocket' like regions (p) of the nuclear envelope. Adjacent to each 'pocket' is a pair of centrioles, one of one pair sectioned transversely (C_1) and one of the other pair sectioned longitudinally (C_2). Clearly stained chromosomes are absent (compare with figs. 48 and 53). × 60,500.

Fig. 39

Detail of a region comparable to that marked X in fig. 38.

Microtubules (sm) from opposite poles of the spindle (sm_1 and sm_2) terminate in kinetochores (arrows). × 79,000.

Figs. 40, 41

Serial cross sections from a comparable area to that marked X in fig. 38. The microtubule arrowed in fig. 40 terminates in a kinetochore (bracketed region), thus it is absent in fig. 41. Use numbered microtubules as references for comparison. × 114,800.

Figs. 38–41 *Glutaraldehyde–osmium tetroxide fixation. From* HEATH, I. B. and GREENWOOD, A. D. (1968). *J. gen. Microbiol.,* **53**, 287.

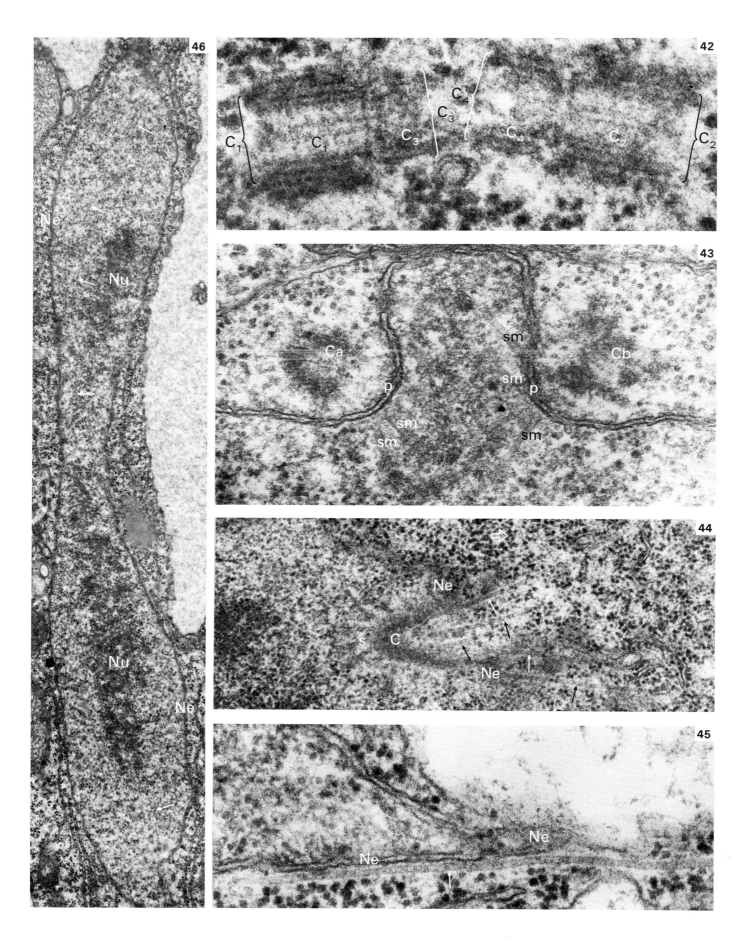

A. Vegetative Structures (cont.)

Mitosis (cont.)

Additional reading

ALDRICH, H. C. (1969). Ultrastructure of mitosis in myxamoebae and plasmodia of *Physarum flavicomum. Am. J. Bot.,* **56**, 290.
HEATH, I. B. and GREENWOOD, A. D. (1970). Centriole replication and nuclear division in *Saprolegnia. J. gen. Microbiol,* **62**, 139.
ICHIDA, A. A. and FULLER, M. S. (1968). Ultrastructure of mitosis in the aquatic fungus *Catenaria anguillulae. Mycologia,* **60**, 141.

Fig. 42

Centriole replication at the initiation of mitosis in *S. ferax.* The two longitudinally sectioned parent centrioles of the interphase pair (C_1 and C_2) have separated and short daughter centrioles (C_3 and C_4) are developing on the proximal end of each parent. $\times 180,200$.

Fig. 43

The next stage after centriole replication (cf. fig. 42) in *S. ferax.* The two pairs of centrioles (Ca and Cb, both sectioned obliquely) are separating and a few spindle microtubules (sm) are developing between the 'pockets' (p). $\times 76,500$.

Fig. 44

One pole of a longitudinally sectioned anaphase/telophase nucleus of *S. ferax* (comparable to fig. 46) showing cytoplasmic microtubules (arrows) radiating into the cytoplasm along the obliquely sectioned nuclear envelope (Ne) from the region of the centriole pair (C). The way in which the nuclear envelope is extended along these microtubules beyond the end of the intranuclear spindle (s) suggests that some force is being generated between the nuclear envelope and the microtubules such that the nucleus moves in the direction of the open arrow. Glutaraldehyde—osmium tetroxide fixation. $\times 48,400$.

Fig. 45

A typical association between the nuclear envelope (Ne) and a cytoplasmic microtubule (arrow). Such a configuration is suggestive of the production of some force which enables the nucleus to 'crawl' along the microtubule. $\times 100,900$.

Fig. 46

An anaphase nucleus of *S. ferax.* The nuclear envelope (Ne) and nucleolus (Nu) are persistent and the spindle (arrows) runs the full length of the nucleus (compare with figs. 49, 54). Glutaraldehyde—osmium tetroxide fixation. $\times 22,300$.

Figs. 42–6 *(42, 43, 45 Glutaraldehyde—osmium tetroxide fixation.) From* HEATH, I. B. *and* GREENWOOD, A. D. *(1970). J. gen. Microbiol.,* **62**, 139. *44 Micrograph by* I. B. HEATH, *York University, Ontario. 46 From* HEATH, I. B. *and* GREENWOOD, A. D. *(1968). J. gen. Microbiol.,* **53**, 287.

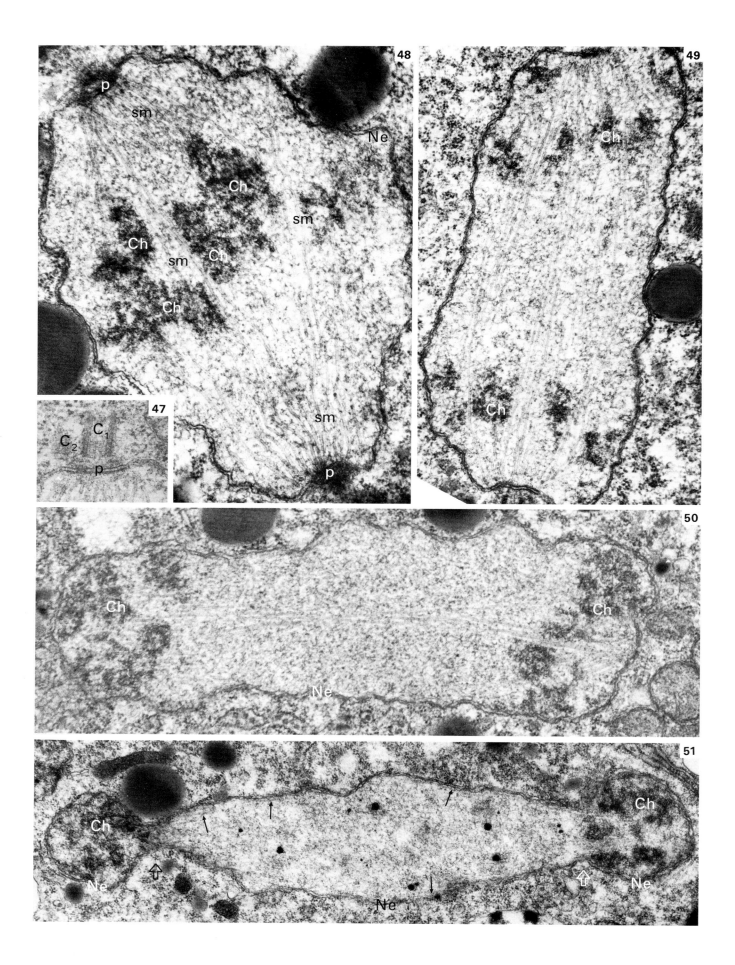

A. Vegetative Structures (cont.)

Mitosis (cont.)

Fig. 47

A detail of one pole of a mitotic spindle of *C. anguillulae* showing the 90° oriented pair of centrioles (C_1 and C_2) adjacent to the 'pocket' (p) of the nuclear envelope. In this class of organisms one centriole (C_1) is typically longer than the other. × 68,700.

Fig. 48

A metaphase nucleus of *C. anguillulae* showing chromosomes (Ch) and spindle microtubules (sm) which converge to the pocket like regions (p) of the nuclear envelope (Ne). × 46,100.

Fig. 49

Anaphase nucleus of *C. anguillulae*. The chromosomes (Ch) are moving towards the poles of the spindle which is elongating in the region between the two chromosome groups. × 32,600.

Fig. 50

Late anaphase nucleus of *C. anguillulae*. The chromosomes (Ch) are now clustered at the poles of the spindle, the nuclear envelope (Ne) is still intact. × 33,900.

Fig. 51

Telophase nucleus of *C. anguillulae*. The two chromosome groups (Ch) are rounding up and becoming 'pinched off' by the nuclear envelope (Ne) from the rest of the nucleoplasm (between open arrows). This part of the nucleus still contains spindle microtubules (arrowed) but will soon break down after the daughter nuclei have been completely detached. × 26,800.

Figs. 47–51 *Glutaraldehyde–osmium tetroxide fixation. From* ICHIDA, A. A. and FULLER, M. S. (1968). *Mycologia,* **60**, 141.

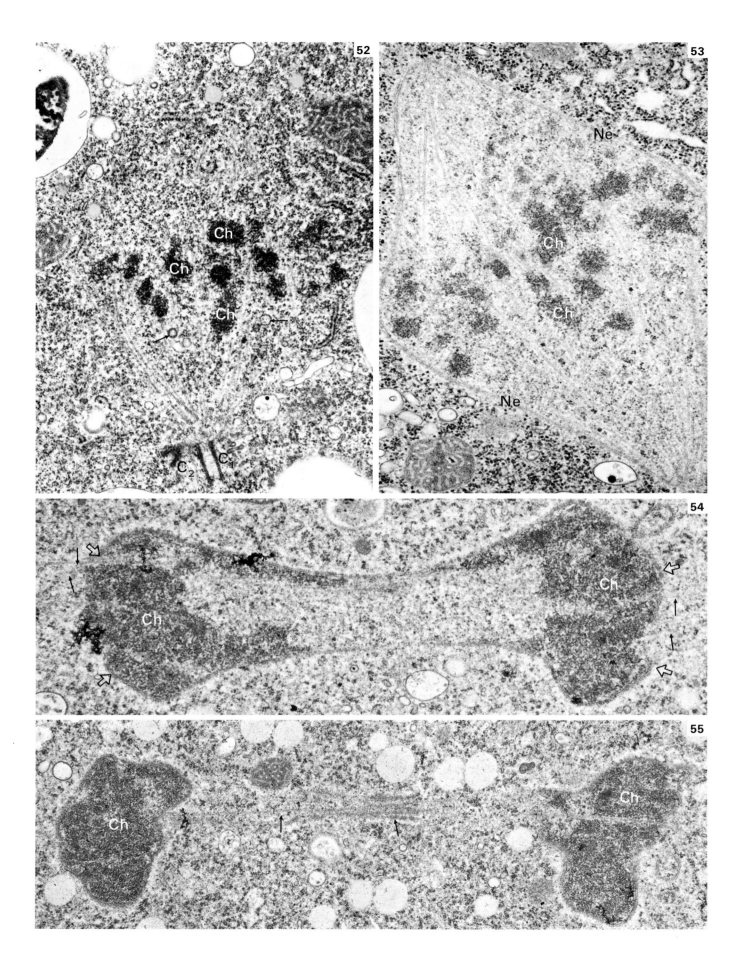

A. Vegetative Structures (cont.)

Mitosis (cont.)

Fig. 52

Metaphase mitotic spindle in a myxamoeba of *P. flavicomum*. One pair of right angle oriented centrioles (C_1 and C_2) are seen· at one pole of the spindle. The plane of section did not intersect those at the other pole. The chromosomes (Ch) are arranged in a well-defined metaphase plate. The nuclear envelope is lacking and vesicles and ribosomes from the cytoplasm have invaded the spindle (arrows). × 13,800.

Fig. 53

Mitotic metaphase spindle in a plasmodium of *P. flavicomum* showing an intact nuclear envelope (Ne), chromosomes (Ch). Note absence of polar centrioles. Compare with fig. 52. × 20,900.

Fig. 54

Late anaphase in a *P. flavicomum* plasmodium nucleus. The chromosomes (Ch) are located at the poles of the spindle, some microtubules of which pass into the cytoplasm (arrows) through polar breaks in the nuclear envelope (between open arrows). × 19,300.

Fig. 55

Telophase plasmodial nucleus of *P. flavicomum*. The nuclear envelope has broken down in the equatorial region of the nucleus and is reforming around the clusters of chromosomes (Ch), thus excluding the spindle (arrows) which soon breaks down. × 16,500.

Figs. 52–5 *Glutaraldehyde–osmium tetroxide fixation. From* ALDRICH, H. C. (1969). *Am. J. Bot.,* **56**, 290.

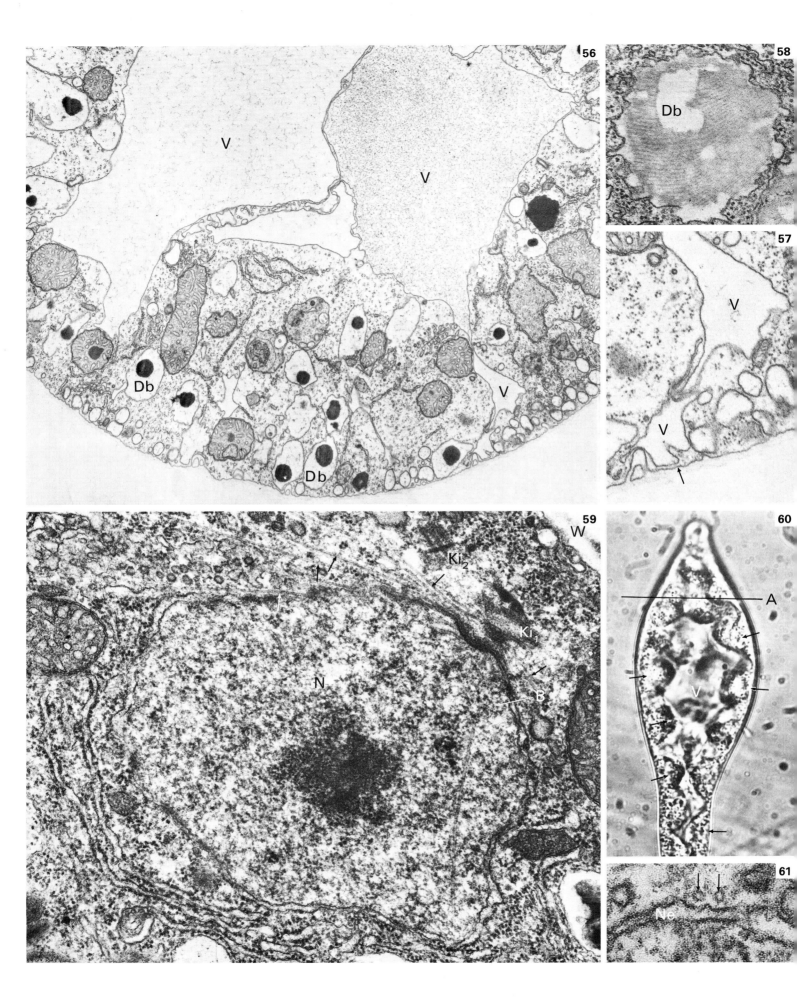

B. Reproductive Structures

Asexual reproduction (zoospore production)

Cytoplasmic cleavage and zoospore formation I. Oömycete

In the biflagellate Oömycetes, for example *Saprolegnia ferax* (Gruithuisen) Thuret, each hyphal tip which is destined to become a zoo sporangium swells and fills with a multinucleate mass of cytoplasm. This sporangium is then isolated from the rest of the hypha by a characteristic **cross wall** (fig. 62). Whilst some nuclei in the sporangium are apparently broken down (fig. 63), most assume a pear shape, each with their apices pointing radially to the cell wall. At the apex of each nucleus the centrioles elongate and reorientate into a 'V' configuration (figs. 59, 64). Since these elongated centrioles subsequently form the basal bodies of the flagella they may now be termed **basal bodies** or **kinetosomes**. Numerous **microtubules** radiate from osmiophilic material around the base of the kinetosomes (figs. 59, 64). Some of these tubules pass into the cytoplasm (fig. 59) whilst others are closely associated with the nuclear envelope (fig. 61). After the above changes have occurred the **central vacuole** of the sporangium enlarges between uninucleate masses of cytoplasm (figs. 56, 60) and ultimately fusion of this vacuole membrane with the cell membrane produces zoospore initials whose membrane is derived in part from the vacuole membrane, and in part from the cell membrane. Enlargement of the central vacuole is probably to some extent due to its fusion with characteristic phospholipid containing **dense bodies** which develop during sporogenesis (figs. 56, 58). It is uncertain what forces are responsible for directing the path of the expanding vacuole so that it always produces uninucleate spore initials, but it is probable that the **microtubules** which radiate around each nucleus (fig. 59) stabilize these areas of cytoplasm and thus the expanding vacuole runs along the line of least resistance, i.e. between the nuclei. (See also Section II, p. 157).

After cytoplasmic cleavage is complete the nine doublet microtubules of the kinetosome elongate to form the characteristic nine doublets of the flagellum **axoneme**. The sheath of the flagellum is formed by expansion of the spore cell membrane. When flagellum elongation is complete, flimmer hairs, which were formed in the endoplasmic reticulum during sporogenesis (figs. 65, 66, see also fig. 82), are attached in two opposite rows to the flagellum sheath (fig. 67). The zoospores develop a water expulsion vacuole (fig. 74) and are then released from the sporangium.

Additional reading

GAY, J. L. and GREENWOOD, A. D. (1966). Structural aspects of zoospore production in *Saprolegnia ferax* with particular reference to the cell and vesicular membranes. In: The fungus spore. *18th Symp. Colston Res. Soc. Bristol.* (M. F. Madelin, ed.), 95. Butterworth, London.

HEATH, I. B. and GREENWOOD, A. D. (1971). Ultrastructural observations on the kinetosomes, and Golgi bodies during the asexual life cycle of *Saprolegnia. Z. Zellforsch. mikrosk. Anat.*, **112**, 371.

Fig. 56

Cross section of a zoosporangium of *S. ferax* at a comparable position to line A in fig. 60. The central vacuoles (V) have expanded between masses of cytoplasm. The dense body vesicles (Db) contain granular material similar to that in the expanding vacuoles, suggesting that they may fuse with the expanding vacuoles, and thus increase their volume. × 19,800.

Fig. 57

Detail of fig. 56 showing the proximity of vacuole (V) and cell membranes (arrow) just prior to fusion and consequent delimitation of zoospores. × 37,000.

Fig. 58

A 'dense body' (Db) showing the typical periodicity which is a common characteristic of phospholipid material. Glutaraldehyde–osmium tetroxide fixation. × 47,600.

Fig. 59

Part of a young zoosporangium of *S. ferax* showing the pyriform nucleus (N) with kinetosomes (Ki_1, Ki_2 obliquely sectioned) adjacent to the cell wall (W). Numerous microtubules (arrows) radiate from the base of the kinetosomes into the cytoplasm and along the surface of the nuclear envelope. Glutaraldehyde–osmium tetroxide fixation. × 36,100.

Fig. 60

Light micrograph of a living sporangium of *S. ferax* at a stage comparable to that shown in fig. 56, showing the enlarging central vacuoles (V) delimiting portions of cytoplasm (arrows) which will become the zoospores. × 1,060.

Fig. 61

Detail of the nuclear envelope (Ne) from region 'B' of a comparable nucleus to that shown in fig. 59 showing the close association between the envelope and microtubules (arrows) (cf. fig. 45). Glutaraldehyde–osmium tetroxide fixation. × 126,400.

Figs. 56–61 (*56, 57 Osmium tetroxide fixation.*) *From* GAY, J. L., GREENWOOD, A. D. and HEATH, I. B. (1971). *J. gen. Microbiol.,* **65**, 233. *58 Micrograph by* I. B. HEATH, York University, Ontario. *59 From* HEATH, I. B. and GREENWOOD, A. D. (1971) 'Ultrastructural observations on the kinetosomes, and golgi bodies during the asexual life cycle of saprogegnia', *Z. Zellforsch. mikrosk. Anat.,* **112**, 371–89. Berlin–Heidelberg–New York: Springer. *60 Micrograph by* DR J. L. GAY, Imperial College, London. *61 Micrograph by* I. B. HEATH, York University, Ontario.

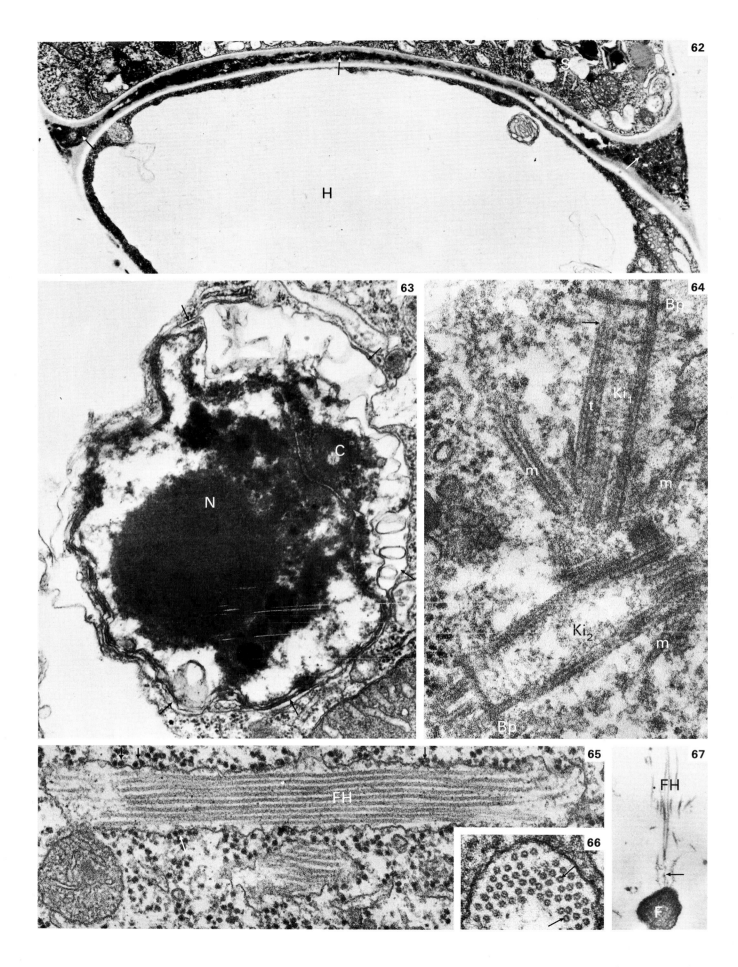

B. Reproductive Structures (cont.)
Asexual reproduction (zoospore production) (cont.)
Cytoplasmic cleavage and zoospore formation I. Oömycete (cont.)

Fig. 62

Longitudinal section of a cross wall at the base of a sporangium (S) of *S. ferax*. This type of wall is inserted within a few seconds and typically contains a layer of trapped cytoplasm (arrows). Note the highly vacuolate hypha (H). × 13,480.

Fig. 63

A nucleus (N) and centriole (C) which have become enveloped in a double membrane bound cisternum (arrows) and are apparently undergoing degeneration. Such a system seems to be a common way of removing excess nuclei from sporangia of *S. ferax*. × 40,300.

Fig. 64

A pair of kinetosomes (Ki$_1$, Ki$_2$) from a comparable sporangium to that shown in fig. 59. They are oriented in a 'V' configuration and have microtubules (m) radiating from their basal region. The third tubule of each of the nine triplet (t) tubules terminates just below the basal plate (Bp) (one such termination is arrowed). Compare these kinetosomes with the centrioles in fig. 42. × 96,000.

Fig. 65

Longitudinally sectioned flimmer hairs (FH) in a ribosome-studded (arrows) cisternum of the endoplasmic reticulum of a pre-cleavage sporangium of *S. ferax*. × 73,700.

Fig. 66

Cross section of flimmer hairs similar to those shown in fig. 65. The tubular nature of these hairs is clearly seen (arrows). × 120,600.

Fig. 67

Cross section of a flagellum (F) of *S. ferax* zoospore showing a part of one of the rows of flimmer hairs (FH) attached to the sheath of the flagellum by typical ill-defined tapering points (arrow). × 32,700.

Figs. 62–7 *Glutaraldehyde–osmium tetroxide fixation.*
62–4 Micrographs by I. B. HEATH, York University, Ontario.
65–7 From HEATH, I. B., GREENWOOD, A. D. and GRIFFITHS, H. B. (1970). *J. Cell Sci.*, **7**, 445.

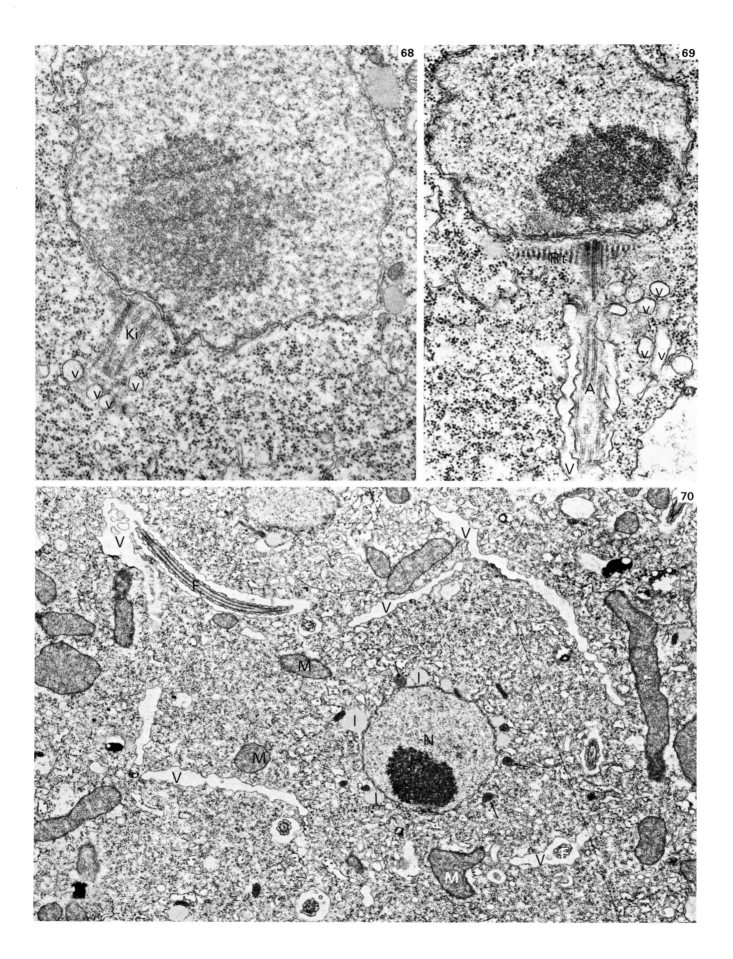

B. Reproductive Structures (cont.)
Asexual reproduction (zoospore production) (cont.)
Cytoplasmic cleavage and zoospore formation II. Chytridiomycete

In a monocentric Chytridiomycete such as *Blastocladiella emersonii* Cantino and Hyatt the whole spherical portion of the thallus becomes separated from the rhizoids by a cross wall and cleaves into zoospores. At the onset of zoosporogenesis one of the two centrioles adjacent to each nucleus elongates into a **kinetosome** (fig. 68). The third tubule of each triplet group of tubules terminates at a comparable length to that in *S. ferax* but the nine doublets continue to elongate as vesicles accumulate and fuse around them in such a way that the **flagellar axoneme** (as these nine doublets now become) grows into a long vacuole (fig. 69). As flagellar elongation continues the flagellum root system (composed in part of a complex striated fibre) matures (fig. 69) and the continued enlargement of the flagella containing vacuoles, and possibly other vacuoles, produces a network of interconnecting **sheet-like vacuoles** (fig. 70)

whose subsequent fusion delimits uninucleate **zoospore initials** (fig. 71). The forces which direct the enlargement of these vacuoles between uninucleate masses of cytoplasm remain unknown.

As cytoplasmic cleavage progresses the **lipid droplets** and **microbodies** congregate around the nucleus (fig. 70). After cleavage the ribosomes of each spore aggregate around the base of the nucleus (fig. 72) and become sequestered from the cytoplasm by a sheet composed of two membranes each of which is continuous with the nuclear envelope (fig. 73, see also fig. 78). When the ribosomes are sequestered in this **cap** they are no longer active in protein synthesis. Prior to release the zoospore matures so that a large single mitochondrion surrounds the base of the flagellum (fig. 73) and the lipid droplets and microbodies form a complex called the **side body** (fig. 73).

Additional reading

LESSIE, P. E. and LOVETT, J. S. (1968). Ultrastructural changes during sporangium formation and zoospore differentiation in *Blastocladiella emersonii*. Am. J. Bot., **55**, 220.

Fig. 68
A nucleus in a sporangium of *B. emersonii* at an early stage of zoospore production. One centriole has elongated to become a kinetosome (Ki) and a few vesicles (v) have clustered around its tip. × 37,000.

Fig. 69
A later stage in cleavage in *B. emersonii*. The flagellar axoneme (A) has elongated and is enclosed by a large vacuole (V) with which smaller vesicles (v) are fusing. The root system (Rt) of the mature spores has developed. × 39,000.

Fig. 70
A sporangium of *B. emersonii* at about the mid point in cleavage. Vacuoles (V), some containing flagella (F), have enlarged and are forming the cleavage furrows around each nucleus (N). Ribosomes and mitochondria (M) are still dispersed but the lipid droplets (I) and microbodies (arrows) are aggregated around the nucleus. × 12,850.

Figs. 68–70 *Glutaraldehyde–osmium tetroxide fixation. Micrographs by* DRS W. E. BARSTOW *and* J. S. LOVETT, *Purdue University, Lafayette, Indiana.*

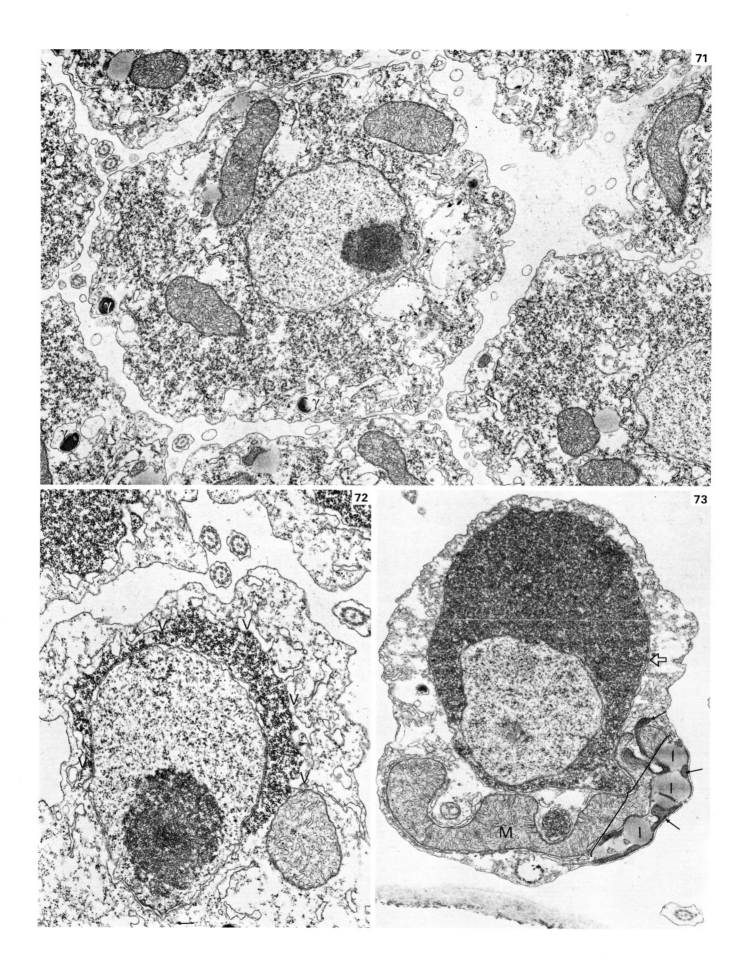

B. Reproductive Structures (cont.)

Asexual reproduction (zoospore production) (cont.)

Cytoplasmic cleavage and zoospore formation II. Chytridiomycete (cont.)

Fig. 71

A late stage in zoosporogenesis of *B. emersonii*. Cytoplasmic cleavage is complete but the ribosomes are still dispersed. Curious structures of unknown function called γ particles are present (γ). ×18,600.

Fig. 72

A maturing zoospore of *B. emersonii* in which the ribosomes are beginning to cluster around the nucleus on the opposite side to the kinetosome (not sectioned, position located by arrow). This developing ribosomal cap is already partially enclosed by vacuoles (V) which may be dilated endoplasmic reticulum. ×24,600.

Fig. 73

A mature zoospore of *B. emersonii*. The ribosomes are enclosed by a pair of membranes (open arrow). One large, lobed mitochondrion is present (M) and the lipid droplets (I) and microbodies (arrows) are aggregated into a side body complex (bracketed). ×18,000.

Figs. 71–3 *Glutaraldehyde–osmium tetroxide fixation.*
Micrographs by DRS W. E. BARSTOW and J. S. LOVETT, Purdue University, Lafayette, Indiana.

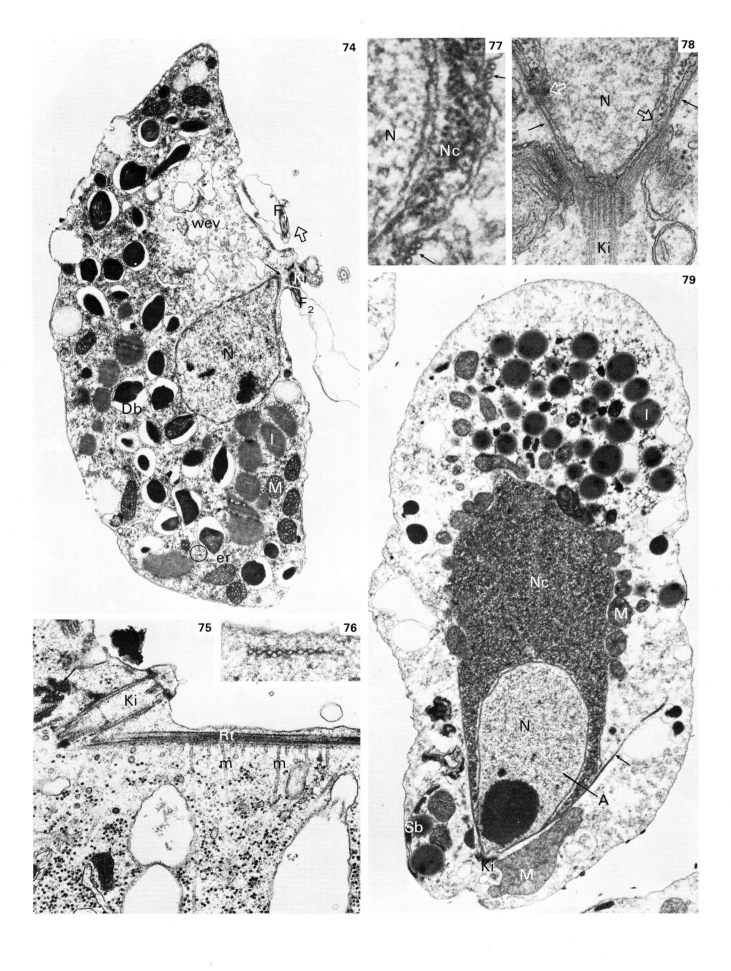

B. Reproductive Structures (cont.)
Asexual reproduction (zoospore production) (cont.)
Zoospore structure

A zoospore has two major design requirements, (a) the need to transport the minimum ingredients for a new colony with a minimum of energy expenditure and, (b) because it uses flagella for propulsion, the need to firmly attach the flagella in such a way as to distribute the reaction to its beating over as much of the spore as possible. By analogy, a whip produces little 'crack' if held in a weak hand. There are a number of degrees of specialization to these ends, of which we shall consider the packaging of the organelles first. In the Oömycetes such as *Saprolegnia ferax*, or *Pythium aphanidermatum* (fig. 74) the zoospore differs relatively little from a uninucleate package of vegetative cytoplasm. Lipid droplets are more abundant and special phospholipid containing 'dense bodies' are formed and packaged into the spore, mainly as energy sources (but see also p. 31) and, as with many wall-less cells, a water expulsion vacuole is required (fig. 74).

The zoospore of the Hyphochytridiomycete *Rhizidiomyces apophysatus* Zopf. (fig. 82) is essentially similar in simplicity to that of the Oömycetes but in this case the ribosomes tend to cluster around the posterior end of the nucleus (fig. 82). Increasingly specialized organization is shown in some members of the Chytridiomycetes, for example *Monoblepharella* sp., where the nuclear cap of ribosomes is more clearly defined but not bounded by membranes (fig. 80), and a specialized structure of unknown function, the **'rumposome'** is developed (fig. 80).

However, the mitochondria remain single and dispersed as in the Oömycetes.

The highest level of complexity of zoospore structure is perhaps found in the Chytridiomycetes such as *Allomyces macrogynus* (Emerson) Emerson and Wilson (fig. 79) and *Blastocladiella emersonii* Cantino and Hyatt (fig. 73). In these zoospores the ribosomes are packaged into a **cap** which is completely enclosed by a membrane which is continuous with the nuclear envelope (figs. 77, 78, 79) and in *B. emersonii* the mitochondria fuse into a single structure around the kinetosome (fig. 73). In both cases lipid droplets and microbodies are aggregated into a special structure termed the **'side body'**, which, like the 'rumposome', is of unknown function (figs. 73, 79). At present the selective advantage which would lead to the evolution of the above zoospore structures remains obscure.

The pattern of a root system which attaches the flagella to the spore is again highly variable through the various fungal groups. Because the nucleus is the largest organelle in the zoospore it is perhaps not surprising that it is frequently the organelle to which the flagella are most firmly attached. In the simplest type of zoospores, the primary zoospores of *S. ferax*, single microtubules radiate from the kinetosomes (in the same way as those shown before cleavage in fig. 59) and either pass into the cytoplasm or are closely associated with the nuclear envelope (figs. 59, 61). In the differently shaped (reniform as opposed to the

Fig. 74

Zoospore of *Pythium aphanidermatum* showing the typical reniform shape with lateral insertion of flagella (F_1, F_2), the anterior one bearing flimmer hairs (open arrow). Single microtubular roots (arrows) run from the kinetosomes (Ki) to the nucleus (N). Ribosomes (circles), dense bodies (Db), lipid droplets (l), mitochondria (M) and endoplasmic reticulum (er) are dispersed throughout the cytoplasm except in the area of the water expulsion vacuole (wev). × 11,500.

Fig. 75

Oblique longitudinal section of the 'backbone' root (Rt) which runs from the kinetosome(s) (Ki) in the spore of *P. aphanidermatum*. Note the laterally attached microtubular 'ribs' (m). The dense material (arrow) is part of the fibre which connects the two kinetosomes. × 75,000.

Fig. 76

Cross section of the 'backbone' root showing the 8 fused microtubules. × 75,000.

Fig. 77

Part of a cross section of an *Allomyces macrogynus* spore at about the position marked A in fig. 79. Two of the nine groups of three microtubules (arrows) are associated with the envelope of the nuclear cap (Nc) adjacent to the nucleus (N). Glutaraldehyde–osmium tetroxide fixation. × 82,800.

Fig. 78

Part of the apex of a longitudinally sectioned nucleus (N) in a zoospore of *A. macrogynus* showing the kinetosome (Ki) and the microtubular roots (arrows) which are associated with the nuclear and cap envelopes. The cap envelope joins the nuclear envelope at the points marked by open arrows. Glutaraldehyde–osmium tetroxide fixation. × 54,000.

Fig. 79

Longitudinal section of a zoospore of *A. macrogynus*. The ribosomes are packaged into a membrane bound cap (Nc) around the anterior of the nucleus (N) and the microtubular roots run from the base of the kinetomes (Ki, sectioned obliquely) along the nuclear envelope and cap envelope into the cytoplasm (arrow). Lipid droplets (l) tend to cluster around the anterior end of the nucleus whilst the mitochondria (M) remain numerous and are also clustered around the nucleus. Part of the side body is seen (Sb). Glutaraldehyde–osmium tetroxide fixation. × 9,850.

Figs. 74–9 (74–6 Glutaraldehyde–osmium tetroxide fixation.) Micrographs by DRS S. N. GROVE and C. E. BRACKER, Purdue University, Lafayette, Indiana. 77 Micrograph by DR M. S. FULLER, University of Georgia, Athens, Georgia. 78 From FULLER, M. S. and CALHOUN, S. A. (1968) 'Microtubule-kinetosome relationships in the motile cells of the Blastocladiales'. *Z. Zellforsch. mikrosk. Anat.*, **87**, 526–33. Berlin–Heidelberg–New York: Springer. 79 Micrograph by DR M. S. FULLER, University of Georgia, Athens, Georgia.

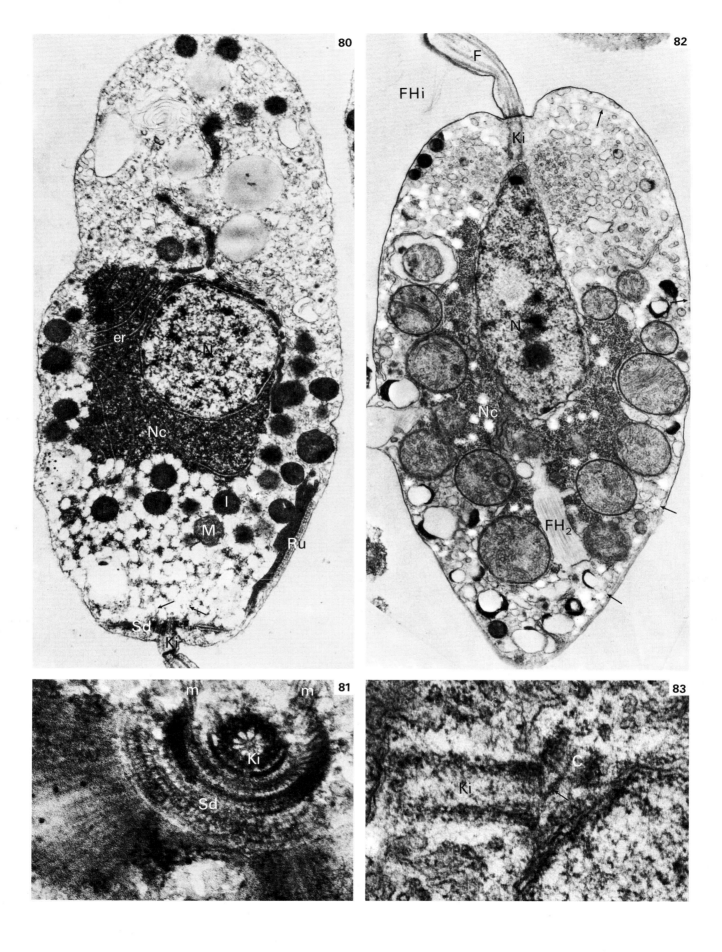

B. Reproductive Structures (cont.)

Asexual reproduction (zoospore production) (cont.)

Zoospore structure (cont.)

pyriform primary spores of *S. ferax*) spores of *Pythium aphanidermatum* some single tubules are also associated with the nuclear envelope (fig. 74), but in addition there is a 'backbone' of 8 tubules fused into a sheet from which 'ribs' of single microtubules radiate (figs. 75, 76).

The root system of *R. apophysatus* is not as clearly described as the above species, but appears to be simple with single microtubular roots which radiate from the kinetosome region and run the length of the spore close to the cell membrane (fig. 82). In this species, as in all the uniflagellate fungi studied to date, only one centriole elongates to form the flagellum, the other is present as a 'vestigial' structure (fig. 83).

In the Blastocladiales, e.g. *Allomyces macrogynus*, the root system is essentially an elaboration of that found in *S. ferax* in that nine 'roots', each composed of three microtubules, run from the base of the kinetosome along the surface of the nuclear envelope and the envelope of the nuclear cap (figs. 79, 77, 78). Again the close association between membranes and microtubules is apparent (cf. figs. 77, 78, 74).

The exceptional root pattern is found in the Monoblepharidales, e.g. *Monoblepharella* sp., where the nucleus is *not* attached to the root system. The base of the kinetosome is set into an elaborate 'striated disk' of osmiophilic material, and from this disk numerous microtubules radiate back through the length of the cytoplasm (figs. 80, 81) but do not contact the nucleus or nuclear cap. The latter structures are set in the middle of the spore as opposed to the more typical position adjacent to the kinetosome.

At present our knowledge of the mechanics of zoospore swimming is too incomplete to allow a clear understanding of the relative merits of these varied systems, but it is interesting to note that the 'simplest' spore, e.g. *S. ferax*, appears to be a much inferior swimmer to the more complex *Allomyces* type.

Additional reading

FULLER, M. S. and CALHOUN, S. A. (1968). Microtubule–kinetosome relationships in the motile cells of the Blastocladiales. *Z. Zellforsch. mikrosk. Anat.*, **87**, 526.

HEATH, I. B. and GREENWOOD, A. D. (1971). Ultrastructural observations on the kinetosomes, and Golgi bodies during the asexual life cycle of *Saprolegnia. Z. Zellforsch. mikrosk. Anat.*, **112**, 371.

REICHLE, R. E. (1969). Fine structure of *Phytophthora parasitica* zoospores. *Mycologia*, **61**, 30.

Fig. 80

Longitudinal section of a zoospore of *Monoblepharella* sp. The ribosomes are arranged in a nuclear cap (Nc) which is not bounded by an envelope but is permeated by endoplasmic reticulum (er). The nucleus (N) lies centrally in the spore whilst the kinetosome (Ki) is embedded in the striated disk (Sd) at the posterior end of the spore. Microtubules radiate into the cytoplasm from this disk (arrows). Lipid droplets (l) and mitochondria (M) are dispersed through the spore. The rumposome (Ru) is seen in the posterior of the spore. × 18,800.

Fig. 81

Surface section of the striated disk (Sd) in a *Monoblepharella* sp. spore showing the kinetosome (Ki) and microtubular roots (m). × 80,400.

Fig. 82

Longitudinal section of a *Rhizidiomyces apophysatus* zoospore. The ribosomes are aggregated into a loose cap (Nc) adjacent to the nucleus (N). The kinetosome (Ki) is adjacent to the nucleus but no roots are visible near the nucleus. The root system is represented by microtubules which run the length of the spore adjacent to the cell membrane (arrows). The flagellum (F) is born anteriorly as are all flagella which bear flimmer hairs (FH$_1$). Other flimmer hairs are also present in a vesicle within the spore (FH$_2$). Osmium tetroxide fixation. × 22,200.

Fig. 83

Pre-cleavage zoospore initial of *R. apophysatus* showing the unusual fibre (arrow) which connects the kinetosome and centriole. Osmium tetroxide fixation. × 63,500.

Figs. 80–3 (80, 81 *Glutaraldehyde–osmium tetroxide fixation.*) *From* FULLER, M. S. and REICHLE, R. E. (1968). *Can. J. Bot.*, **46**, 279. *82 From* FULLER, M. S. and REICHLE, R. E. (1965). *Mycologia*, **57**, 946. *83 From* FULLER, M. S. (1966). The fungus spore, *18th Symp. Colston Res. Soc. Bristol* (M. F. Madelin, ed.), 67. Butterworth, London. Reprinted by permission of the Colston Research Society.

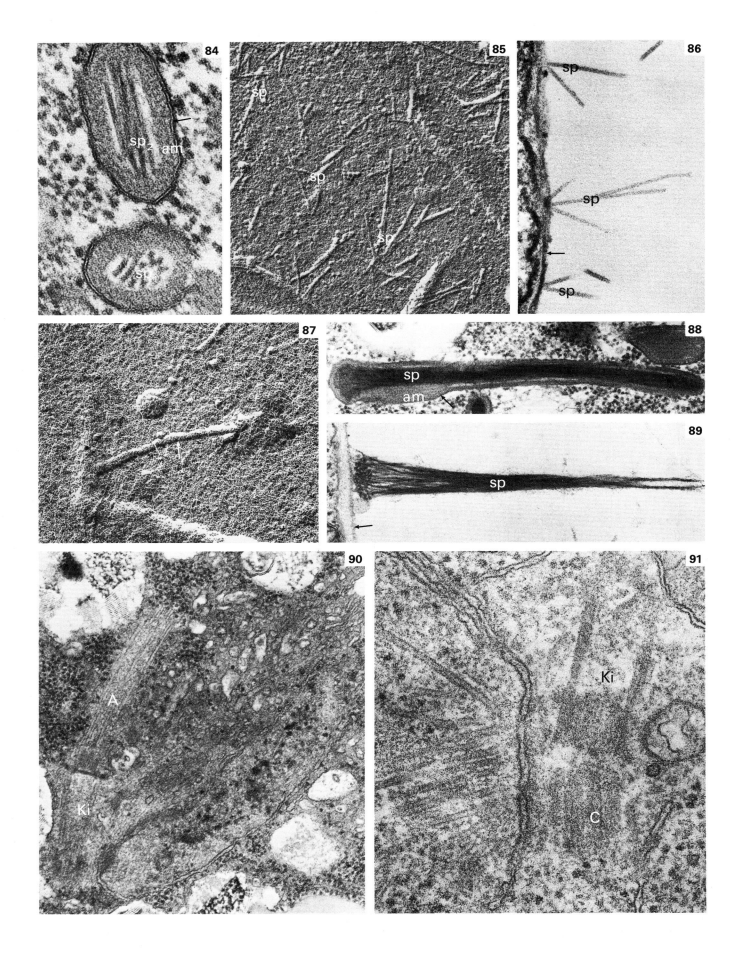

B. Reproductive Structures (cont.)

Asexual reproduction (zoospore production) (cont.)
Zoospore encystment

At the end of its motile life a zoospore forms a spherical, walled resting spore, or cyst. During encystment, depending upon the type of spore, the flagella may be shed at a level just above the basal plate or the axoneme may be withdrawn (fig. 90) leaving the flimmer hairs, if present, attached to the exterior of the cyst. Once inside the cyst the axoneme is usually rapidly broken down to the kinetosomes which, at least in *Saprolegnia ferax*, revert to centrioles at the time of mitosis during cyst germination (fig. 91).

In *S. ferax* the outer, first formed, layer of the cyst wall is produced by the fusion of membrane bound vesicles, or 'bars' (figs. 84, 88) with the cell membrane and the con-

sequent release of their contents to the cell surface. These 'bars', or functionally similar structures in other species, contain amorphous material which forms the base of the wall layer, and elaborate spines (figs. 84, 88) which adorn the cyst surface (figs. 85, 86, 87, 89) and whose morphology appears to be specific to both species (figs. 85, 89) and spore types within a species (figs. 85, 87). The 'bars' are formed at an early stage in spore cleavage, are carried in the motile spore and finally released. The formation of the outer wall layer is very rapid, subsequently the wall is thickened by the addition of an inner wall layer derived in part from Golgi vesicles (cf. 'hyphal growth', p. 5).

Additional reading

HEATH, I. B. and GREENWOOD, A. D. (1971). Wall formation in the Saprolegniales II. Formation of cysts by the zoospores of *Saprolegnia* and *Dictyuchus. Arch. Mikrobiol.*, **75**, 67.

Fig. 84
Two 'bars' in a primary zoospore of *S. ferax* showing the membrane (arrow), a layer of amorphous material (am) and a group of tubular spines sectioned transversely (sp_1) and longitudinally (sp_2). ×115,500.

Fig. 85
Replica of the outer surface of a primary cyst (from a primary zoospore) of *S. ferax* showing clusters of spines (sp), of a similar size to those shown in fig. 84, on an amorphous background. Pd/Au shadowed. ×47,400.

Fig. 86
Section of the surface of a primary cyst of *S. ferax* showing the spines (sp) mounted on an osmiophilic thin outer wall layer (arrow). ×83,500.

Fig. 87
Replica of the outer surface of a secondary cyst of *S. ferax* showing a typical 'boathook' shaped spine (arrow) characteristic of this spore type (cf. fig. 85). Pd/Au shadowed. ×14,100.

Fig. 88
The equivalent of a 'bar' in a primary cyst of *Dictyuchus*

sterile Coker showing a large spine (sp) surrounded by amorphous material (am) and a membrane (arrow). ×43,100.

Fig. 89
Section of a secondary cyst (cyst produced by the zoospore which was released from the primary cyst) of *D. sterile* showing a spine (sp) attached to the osmiophilic outer wall layer (arrow). ×45,300.

Fig. 90
A primary zoospore of *S. ferax* fixed within seconds of encysting. The flagellar axoneme (A), attached to the kinetosome (Ki), has been withdrawn. ×30,700.

Fig. 91
A residual kinetosome (Ki) and its associated centriole (C) at the pole of a second mitotic division spindle in a germinating primary cyst of *S. ferax*. ×91,800.

Figs. 84–91 (84, 86, 88–91 Glutaraldehyde–osmium tetroxide fixation.) 84–89 From HEATH, I. B. and GREENWOOD, A. D. (1971). *Arch. Mikrobiol.*, **75**, 67. 90, 91 From HEATH, I. B. and GREENWOOD, A. D. (1971) 'Ultrastructural observations on the kinetosomes, and golgi bodies during the asexual life cycle of saprolegnia', *Z. Zellforsch. mikrosk. Anat.*, **112**, 371–89. Berlin–Heidelberg–New York: Springer.

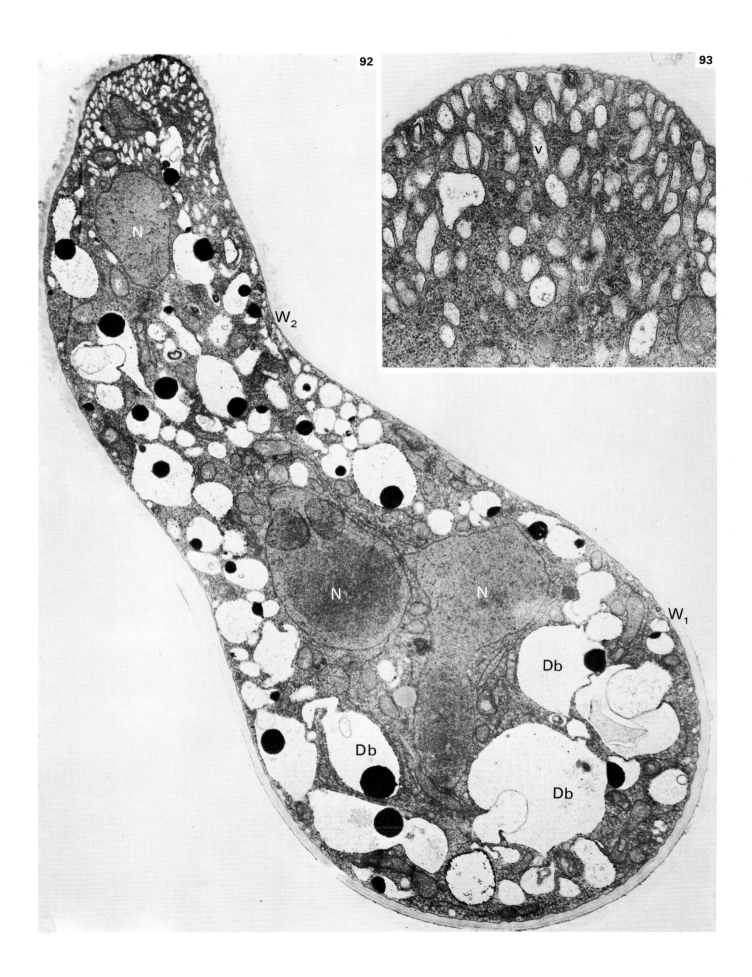

92

93

B. Reproductive Structures (cont.)
Asexual reproduction (zoospore production) (cont.)
Cyst germination

The final stage in the asexual life cycle is the germination of the cyst to produce a germ tube and thus a new vegetative colony. The protrusion of the germ tube in some Oömycetes, e.g. *S. ferax*, is preceded by a localized accumulation of wall vesicles adjacent to a portion of the cyst wall. The osmiophilic outer cyst wall is eventually ruptured and the germ tube grows out as an extension of the inner layer of the cyst wall (fig. 92). From its earliest stage the tip of the germ tube is filled with wall vesicles (fig. 93) and is morphologically similar to a vegetative hypha (cf. fig. 1). As the germ tube emerges mitosis begins, thus producing the coenocytic state of the mature colony.

Since the wall vesicles accumulate prior to cyst germination, it is possible that they contain enzymes responsible for softening the cyst wall and thus allowing the germ tube to be pushed out. However, pressure must be produced within the cyst in order to expand the softened wall. This pressure is apparently produced by expansion of the 'dense body' vesicles (fig. 92) whose subsequent enlargement and fusion produces the central vacuole of the germling.

Additional reading

GAY, J. L., GREENWOOD, A. D. and HEATH, I. B. (1971). The formation and behaviour of vacuoles (vesicles) during oösphere development and zoospore germination in *Saprolegnia*. *J. gen. Microbiol.*, **65**, 233.

Fig. 92

Germinating cyst of *S. ferax* showing the extension of the inner cyst wall (W$_1$) to form the germ tube wall (W$_2$). The germ tube apex is filled with wall vesicles (v) and the 'dense body' vesicles (Db) have enlarged considerably. At least two mitoses have occurred in the spore since there are three nuclei (N) in this section. Glutaraldehyde–osmium tetroxide fixation. × 9,800.

Fig. 93

Apex of a comparable germ tube to that shown in fig. 92. Wall vesicles (v) are abundant. Glutaraldehyde–osmium tetroxide fixation. × 30,900.

Figs. 92–3 *92 From* GAY, J. L., GREENWOOD, A. D. and HEATH, I. B. (1971). *J. gen. Microbiol.*, **65**, 233. *93 Micrograph by* I. B. HEATH, York University, Ontario.

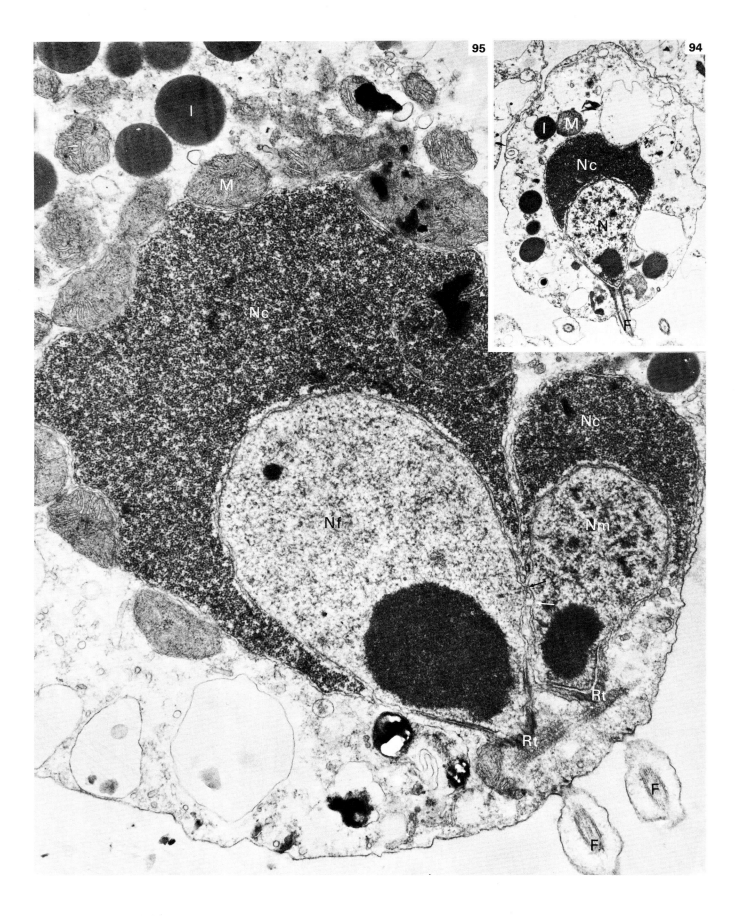

95

94

B. Reproductive Structures (cont.)
Sexual reproduction
Gamete fusion I. Allomyces

One of the types of sexual reproduction common in the flagellate fungi, especially the Chytridiomycetes, is the production, and subsequent fusion, of flagellate, free swimming gametes. For example, in *Allomyces macrogynus* (Emerson) Emerson and Wilson, morphologically distinct **anisogametes** are produced in separate **gametangia**. Both gametes are similar to the zoospores in general morphology (compare figs. 94 and 79), but the female is considerably larger, the nucleus and **nuclear cap** are bigger and there are more mitochondria and lipid droplets present than in the male (fig. 95). The gametes are released from the gametangia, the males are attracted to the females by a hormone, sirenin, and subsequently they fuse in appropriate pairs (fig. 95). After fusion of the cytoplasm the nuclear caps and nuclei fuse to produce a uninucleate, biflagellate **zygote** which subsequently encysts and eventually germinates to form the diploid sporophyte generation of the life cycle.

Fig. 94

Male gamete of *Allomyces macrogynus* showing the typical nuclear cap (Nc), nucleus (N) and single posterior flagellum (F). Mitochondria (M) and lipid droplets (I) are sparse. × 17,100.

Fig. 95

Zygote of *A. macrogynus* shortly after gamete fusion. The nuclear cap (Nc) and nucleus (N) of the female (suffix f) are larger than those of the male (suffix m) and most of the mitochondria (M) and lipid droplets (I) are from the female. The nuclear envelopes have just begun to fuse (arrows). The nuclear caps may be fusing in another region of the spore, if not they will shortly do so. The zygote clearly possess two flagella (F) and root systems (Rt) around the bases of the kinetosomes (which are out of the plane of section). × 30,800.

Figs. 94, 95 *Glutaraldehyde–osmium tetroxide fixation. Micrographs by* DR M. S. FULLER, *University of Georgia, Athens, Georgia.*

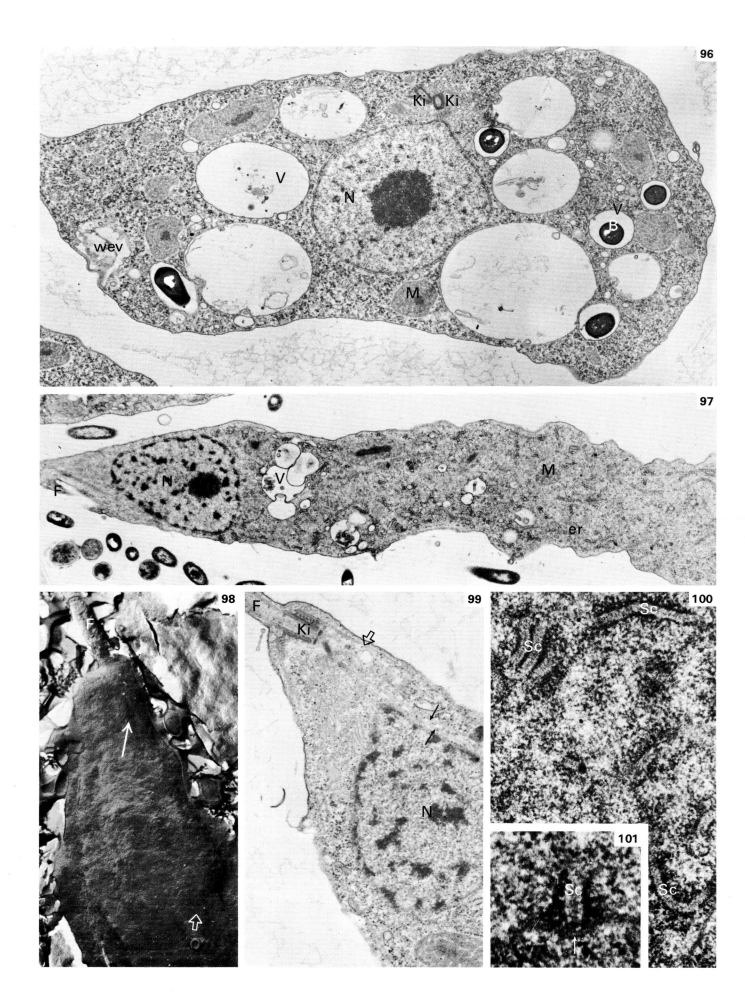

B. Reproductive Structures (cont.)
Sexual reproduction (cont.)
Gamete fusion II. Myxomycetes

Gamete production in the Myxomycetes is a more complex process than in *Allomyces* species. The diploid plasmodium produces a sporangium in which tough, resistant uninucleate spores are formed. Meiosis, as witnessed by the presence of **synaptonemal complexes** (see also Section II, p. 149), occurs in these spores (figs. 100, 101) and subsequently they germinate to produce a haploid **myxamoeba** or **flagellated swarm cell** (figs. 96, 97). Myxamoebae are uninucleate, with a pair of kinetosomes (fig. 96), and, under suitable conditions, may live an independent existence, feeding by pinocytosis and subsequent digestion (fig. 96), as in the plasmodium, and multiplying by mitosis and cell division (see p. 23). Ultimately they may either fuse with another myxamoeba and develop into a new diploid, coenocytic plasmodium, or they may transform into a flagellated swarm cell.

A swarm cell typically has two flagella, one functional whiplash flagellum (figs. 98, 99) and a much shorter one which is closely appressed to the body of the spore and is non-functional in swimming (fig. 97). The kinetosome of the functional flagellum typically has two sets of roots, one of single microtubules which are associated with the nucleus and another set which run along the spore near its surface (fig. 99). These microtubular roots keep the apex of the swarm cell smooth and apparently rigid (figs. 98, 99) whilst the other end of the cell is amoeboid and may feed pinocytotically. Swarm cells are free living but do not divide. They may either fuse with each other in pairs and then develop into a diploid plasmodium or revert to myxamoebae which can subsequently fuse and produce plasmodia.

Additional reading

ALDRICH, H. C. (1967). Ultrastructure of meiosis in three species of *Physarum*. Mycologia, **59**, 127.
ALDRICH, H. C. (1968). The development of flagella in swarm cells of the myxomycete *Physarum flavicomum*. J. gen. Microbiol., **50**, 217.

Fig. 96
Myxamoeba of *Physarum flavicomum* Berk. showing the nucleus (N) with two kinetosomes (Ki), numerous mitochondria (M), food vacuoles (V) with undigested bacteria (B) and a water expulsion vacuole (wev). × 10,600.

Fig. 97
Swarm cell of *Physarella oblonga* (Berk. and Curt.) Morgan showing the apical nucleus (N), the short flagellum (F), a food vacuole (V), mitochondria (M) and in the amoeboid end of the cell, endoplasmic reticulum (er). × 10,300.

Fig. 98
Freeze-etched replica of the surface of a swarm cell of *P. flavicomum* showing the smooth apical cell membrane (arrow), more irregular amoeboid part (open arrow) and flagellar membrane (F). × 18,000.

Fig. 99
Apical region of a *P. oblonga* swarm cell showing flagellum (F) kinetosome (Ki) and the two sets of microtubular roots. Those along the cell surface (open arrow), and those adjacent to the nucleus (N) (arrows). × 24,300.

Fig. 100
Synaptonemal complexes (Sc) in a meiotic prophase nucleus of a spore of *P. flavicomum*. × 29,500.

Fig. 101
Part of the same nucleus as shown in fig. 100 showing the attachment of one end of a synaptonemal complex (Sc) to the nuclear envelope (arrow). × 79,800.

Figs. 96–101 (96, 97, 99–101 Glutaraldehyde–osmium tetroxide fixation.) 96–99 Micrographs by DR H. C. ALDRICH, University of Florida, Gainsville, Florida. 100–1 Micrographs by DRS P. B. MOENS, York University, Ontario, and H. C. ALDRICH, University of Florida, Gainsville, Florida.

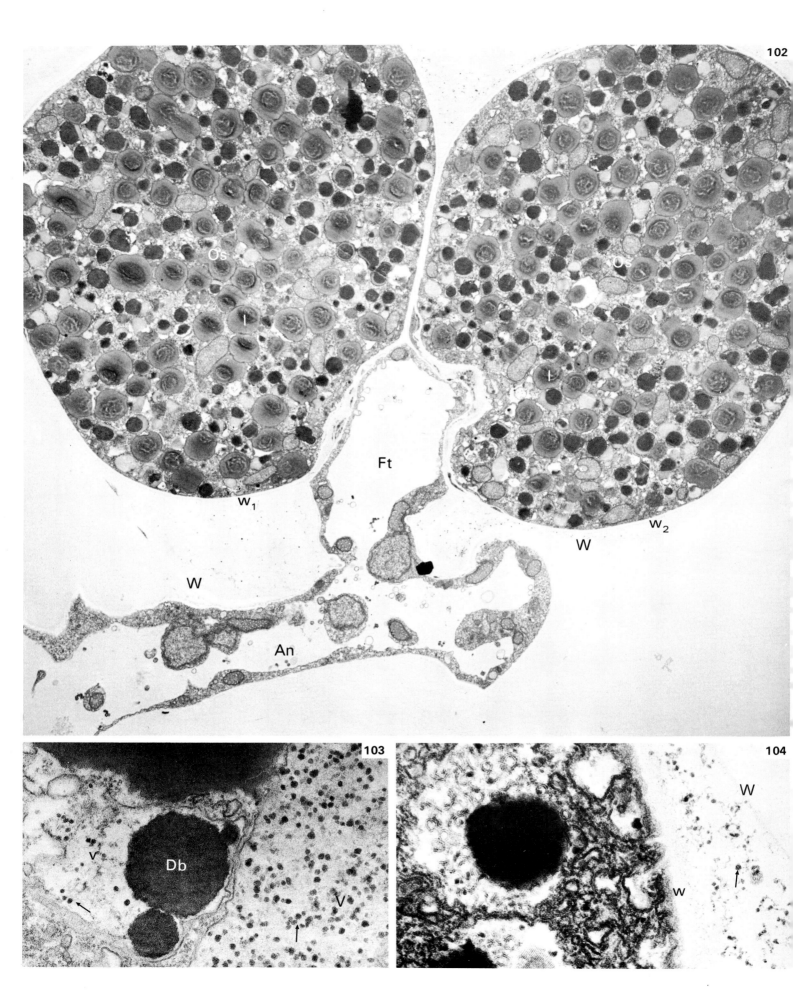

B. Reproductive Structures (cont.)
Sexual reproduction (cont.)
Gametangial fusion

In many fungi, such as the Oömycetes, sexual reproduction involves the fusion of non-motile, differentiated hyphal branches and the subsequent transfer and fusion of nuclei. In *Saprolegnia* certain hyphal side branches swell into spherical **oögonia** which are then cut off from the rest of the colony by cross walls like those of the sporangia (see fig. 62). The nuclei divide **meiotically** (figs. 105, 106, 108) and most of the resulting haploid nuclei degenerate so that when cytoplasmic cleavage ensues uninucleate **oöspheres** are produced (fig. 102). Cytoplasmic cleavage is achieved by enlargement of the central vacuole in a similar way to cleavage in the zoosporangia of this genus (see p. 31). Enlargement of the central, cleaving vacuole is brought about in part by the **dense body vesicles** which

fuse with it (figs. 103, 104). After cleavage each oösphere secretes a cell wall about itself.

Concurrently with the above process the antheridial branch has grown into a multibranched, polymorphic **antheridium** in which the nuclei also undergo meiosis. Branches of the antheridium then penetrate the oögonium wall (fig. 102) and **fertilization tubes** grow to contact the oöspheres (figs. 102, 107). One fertilization tube breaches the oösphere wall and an antheridial nucleus enters the oösphere (fig. 107). Subsequently nuclear fusion occurs to produce the uninucleate, diploid zygote or **oöspore**, a highly resistant resting spore which will subsequently germinate into a new colony.

Additional reading

HOWARD, K. L. and MOORE, R. T. (1970). Ultrastructure of oögenesis in *Saprolegnia terrestris. Bot. Gaz.*, **131**, 311.
GAY, J. L., GREENWOOD, A. D. and HEATH, I. B. (1971). The formation and behaviour of vacuoles (vesicles) during oösphere development and zoospore germination in *Saprolegnia. J. gen. Microbiol.*, **65**, 233.

Fig. 102

An oögonium of *Saprolegnia furcata* Maurizia showing two oöspheres (Os) which typically contain abundant lipid droplets (l), surrounded by oösphere walls (w_1, w_2). An antheridium (An) is applied to the oögonium, a fertilization tube (Ft) has penetrated the oögonium wall (W) and is attached to the oöspheres. Glutaraldehyde—osmium tetroxide fixation. × 7,500.

Fig. 103

Part of a pre-cleavage oögonium of *S. furcata* showing dense bodies (Db). The dense body vesicles (v) and the enlarging central vacuole (V) have similar contents (arrows). × 62,000.

Fig. 104

A post cleavage oögonium of *S. furcata*. The characteristic granules, once present in the central vacuole (arrow, and fig. 103), are now found in the space between the oösphere wall (w) and the oögonium wall (W). This may be evidence that the cleavage process results in the release of the vacuole contents to this space, as would occur when the vacuole and cell membrane fused to complete the cleavage process. × 58,000.

Figs. 102–4 *(103–4 Osmium tetroxide fixation.) 102 Micrograph by* DR J. L. GAY, *Imperial College, London. 103–4 From* GAY, J. L., GREENWOOD, A. D. and HEATH, I. B. (1971). *J. gen. Microbiol.*, **65**, 233.

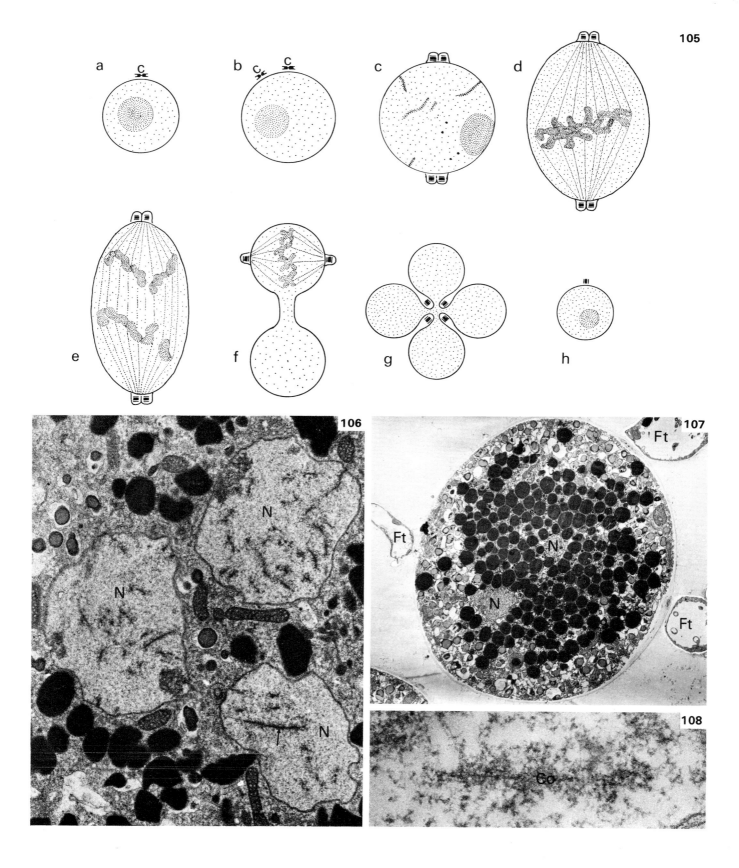

105

a   c

b   c   c

c

d

e

f

g

h

106

N

N

N

107

Ft

Ft

N

N

Ft

108

Co

B. Reproductive Structures (cont.)

Sexual reproduction (cont.)

Gametangial fusion (cont.)

Fig. 105

Diagram depicting meiosis in *Saprolegnia terrestris* Cookson ex Seymour. The centrioles (C) of the interphase nucleus, 105 a, replicate, 105 b, and migrate to opposite sides of the nucleus, 105 c. Axial cores are found in the nucleus, suggesting that meiosis is occurring, 105 c. A spindle develops within the persistent nuclear envelope, 105 d, and chromosome separation (anaphase) of Meiosis I occurs, 105 e. *Without separation of the original nuclear envelope* the pairs of centrioles migrate, without replicating, so that at the end of each Meiosis II spindle (only one shown in 105 f) there is only one centriole. Chromosome separation occurs, a curious cloverleaf configuration is assumed, 105 g, and finally four haploid nuclei are formed, each with only one centriole, 105 h.

Fig. 106

Three prophase meiotic nuclei (N) of *S. terrestris* showing the axial cores (arrow). × 12,500.

Fig. 107

Fertilized oösphere of *S. terrestris* showing the two haploid nuclei (N) prior to fusion. Note the fertilization tubes (Ft). × 4,500.

Fig. 108

Detail of a lateral component or axial core (Co). In most meiotic systems homologous chromosomes are held together by the pairing of two lateral components in a synaptonemal complex, the structure apparently required for genetic crossing over to occur. To date no synaptonemal complex has been seen in any Oömycete (see however fig. 100 and Section II, fig. 294). × 42,000.

Figs. 105–8 *(106–8 Glutaraldehyde/acrolein–osmium tetroxide fixation.) From* HOWARD, K. L. *and* MOORE, R. T. (1970). *Bot. Gaz.,* **131**, 311. © By The University of Chicago. All Rights Reserved.

SECTION 1b

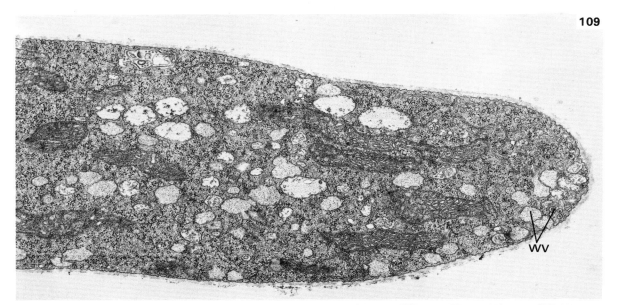

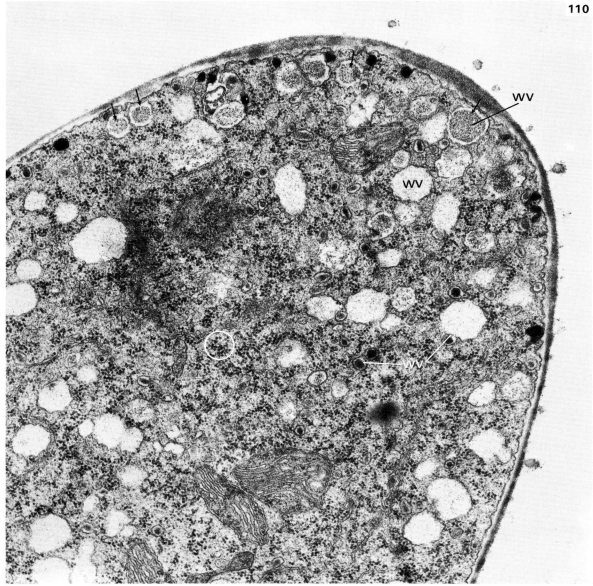

A. Vegetative Structures

Hyphae I. Cell wall synthesis

Cell wall structure in the vegetative hyphae of Zygomyco-tina resembles that already described for the flagellate Oömycete fungi (p. 5) in that a network of **fibrils** is formed embedded in an **amorphous matrix**. However, in Zygomycotina the fibrils are made up of chitin.

As with most fungi, wall synthesis at the hyphal tip is associated with the formation of **wall vesicles** which accumulate at the apex and which apparently become incorporated into the developing cell wall.

In *Rhizopus sexualis* (Smith) Callen wall vesicles are of two distinct types (fig. 110). Large ones with granular contents of varying electron-density and small ones with electron-dense contents. Structural continuities between wall vesicle membrane and the cell plasma membrane suggest that these membranes fuse with each other so enabling the vesicle contents to be deposited into the cell wall. The chemical composition of the contents of wall vesicles is unknown, but as explained for Oömycetes (p. 5), selective staining suggests that **polysaccharide** is the major component.

The sites of wall vesicle formation are located behind the growing tip in the sub-apical region of the hypha. However, unlike *Pythium* hyphae (figs. 1, 2), the hyphae of *R. sexualis* do not possess the typical **stacked cisternae** of the **Golgi dictyosomes**, but contain what are probably functional equivalents of the Golgi dictyosomes in the form of ring-like cisternae which may be termed **Golgi cisternae** (fig. 114, see also Section II, p. 85; Section III, p. 191). It is from these Golgi cisternae that wall vesicles are thought to be formed by a process of budding or blebbing.

Additional reading

GROVE, S. N. and BRACKER, C. E. (1970). Protoplasmic organization of hyphal tips among fungi: vesicles and Spitzenkörper. *J. Bact.*, **104**, 989.

SYROP, MARY (1973). The ultrastructure of the growing regions of aerial hyphae of *Rhizopus sexualis* (Smith) Callen. *Protoplasma*, **76**, 309.

Fig. 109

Median longitudinal section through the tip and sub-apical region of an aerial hypha of *Rhizopus sexualis* (Smith) Callen. Two types of wall vesicles can be seen (wv) together with several elongate mitochondria which are typically orientated parallel to the long axis of the hypha. Glutaraldehyde/formalde-hyde—osmium tetoxide fixation. × 14,200.

Fig. 110

Median longitudinal section through the tip of an aerial hypha of *Rhizopus sexualis*, showing the accumulation of large and small wall vesicles (wv). The large vesicles near the plasma membrane at the tip of the hypha (arrows) contain granular material which is apparently more concentrated than that in similar large vesicles further back from the apex. This may be an artefact of fixation but it may also represent a concentration or a change in composition of material within the vesicles prior to it becoming incorporated within the wall. Ribosomes which are abundant in the sub-apical region (circled) are absent from the extreme apical zone. Glutaraldehyde/formaldehyde—osmium tetroxide fixation. × 37,000.

Figs. 109, 110 *From* SYROP, MARY (1973). *Protoplasma*, **76**, 309.

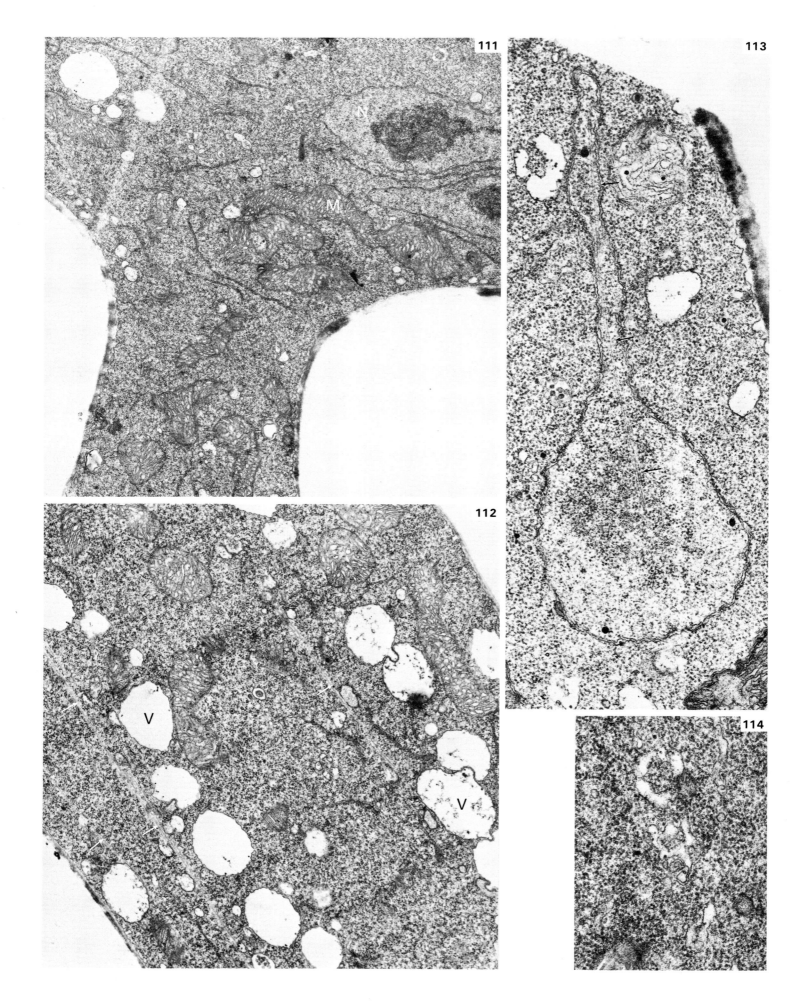

A. Vegetative Structures (cont.)

Hyphae II. Cell components

Mitochondria are numerous in the sub-apical region of vegetative hyphae. The cristae are plate-like after $KMnO_4$ fixation, but when fixed with aldehydes they are tubular, irregularly arranged and usually numerous (figs. 111, 112). All normal mitochondria which have been suitably fixed with an aldehyde–osmium tetroxide fixative contain ribosomes and DNA fibrils within their matrices. As in Oömycetes (p. 13), mitochondria are small near the hyphal apex or in young branches, while those in older parts are considerably larger.

As in most fungi, interphase **nuclei** occur some distance behind the hyphal apex, are bound by a perforated double membrane, the **nuclear envelope** and possess a dense, irregular **nucleolus** (fig. 111).

Profiles of the **endoplasmic reticulum** membranes are abundant and the cytoplasm is densely packed with ribosomes (figs. 111, 112). As stated previously (p. 57), typical **Golgi dictyosomes** have not been found in hyphae of Zygomycotina. However irregular, single **Golgi cisternae** (fig. 114) are numerous and are often seen associated with discrete vesicles, or apparently in the process of budding to produce vesicles. (See also Section II A; fig. 157, Section III A; fig. 365).

Microtubules are often found in Zygomycotina hyphae (fig. 112). Their function remains speculative but they may be involved in organelle motility as previously described (p. 13).

Vacuolization of hyphal cells increases with age often involving the fusion of many small **vacuoles** to form larger, irregular ones. Vacuoles may serve as containers of reserve food materials or depositories for unwanted or even harmful products.

Additional reading

BRACKER, C. E. (1967). Ultrastructure of fungi: *A. Rev. Phytopathol.*, **5**, 343.
MORRÉ, D. J., MOLLENHAUER, H. H. and BRACKER, C. E. (1971). Origin and continuity of Golgi apparatus. In: *Results and Problems in Cell Differentiation II. Origin and Continuity of Cell Organelles* (T. Reinert and H. Ursprung, eds.), pp. 82–126. Springer-Verlag, Berlin.

Fig. 111

Longitudinal section through part of a branching hypha of *Rhizopus sexualis* (Smith) Callen showing interphase nuclei (N) with prominent nucleoli, large, multicristate mitochondria (M), endoplasmic reticulum membranes and the cytoplasm densely packed with ribosomes. Several small vesicles and vacuoles are also present. × 13,600.

Fig. 112

Longitudinal section through part of the sub-apical region of a hypha of *R. sexualis* showing several cytoplasmic microtubules (arrows) lying parallel to the long axis of the hypha. Several vacuoles (V) are also seen apparently lined up along the microtubules. × 17,000.

Fig. 113

A nucleus of *R. sexualis* with a narrow extension of the nuclear envelope within which lies a microtubule (arrows). This unusual feature may indicate that the nucleus had recently divided and the microtubule may be a relic of the division spindle. Serial sections of this nucleus confirmed that the extension was only a very narrow one and a corresponding daughter nucleus was not observed. × 37,000.

Fig. 114

Golgi cisternae of *R. sexualis*. These cisternae produce smooth membrane-bound vesicles which may be involved in secretion and transport of materials during hyphal wall synthesis. × 48,000.

Figs. 111–14 *Glutaraldehyde/formaldehyde–osmium tetroxide fixation. 111, 113–14 Micrographs by* MARY SYROP, Department of Botany, Bristol University. *112 From* SYROP, MARY (1973). *Protoplasma*, **76**, 309.

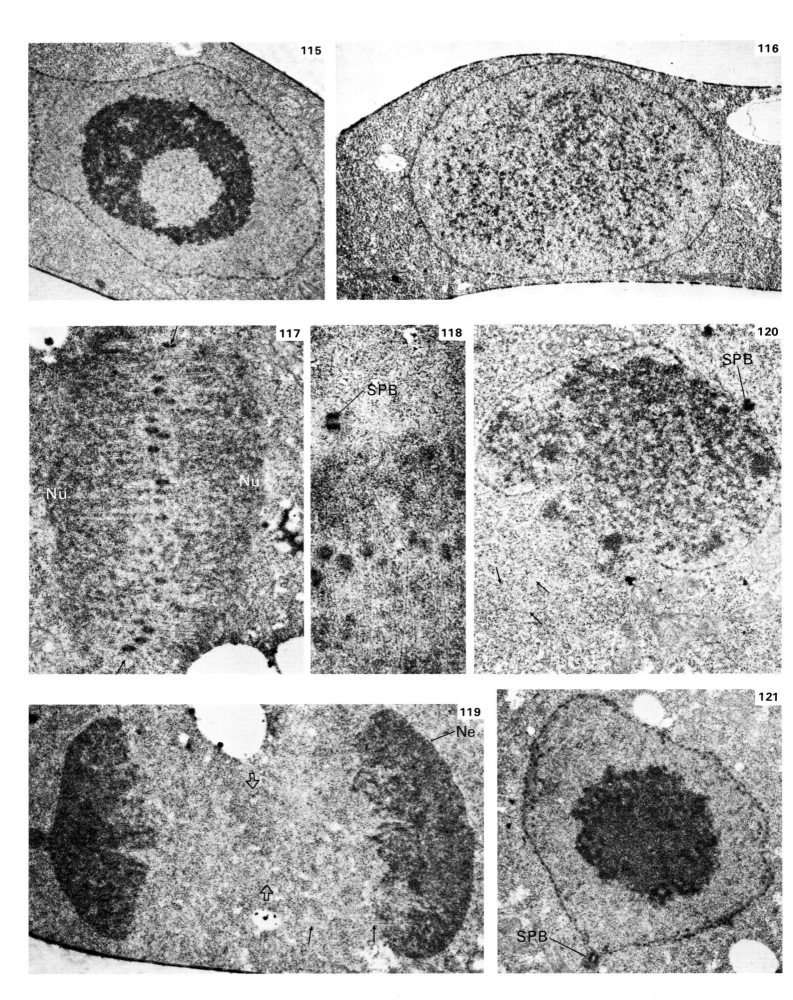

A. Vegetative Structures (cont.)

Mitosis

Very little is known of the ultrastructural details of mitosis in zygomycetous fungi. The fungus *Basidiobolus ranarum* Eidam provides an exception to the above statement since it has been the subject of several extensive studies. However, it is in many respects atypical of Zygomycotina.

The large nucleus of *B. ranarum* (*ca.* 15–20 μm × 5·0 μm) is readily detectable both with the light and electron microscope, and so correlative techniques have been applied to aid our understanding of mitosis in this fungus. The interphase nucleus possesses a large heterogeneous nucleolus (fig. 115). During prophase the nucleolar material becomes dispersed but remains enclosed by the intact nuclear envelope (fig. 116). However, it has been shown that during late prophase the nuclear envelope begins to break down and cytoplasmic microtubules form in large numbers around the nucleus. These microtubules later become associated with small, electron-opaque granular structures which are interpreted as being chromosomes. At metaphase the chromosomes become aligned along a distinct **metaphase plate**, a feature which is uncommon in fungal mitosis (see Section I, p. 23) Section II, p. 93; Section III, p. 199). Large numbers of **spindle microtubules** can now be found both associated with chromosomes and traversing the metaphase plate (figs. 117, 118). Diffuse nucleolar material occurs on either side of the metaphase plate and marks the poles of the wide barrel-shaped spindle (fig. 117). An electron-opaque, tubular body of unknown composition and structure is associated with the poles of the spindle at this stage (fig. 118). This organelle, the **spindle pole body**[1] may either lie 'end on' or 'side on' (fig. 118), to the metaphase plate. Separation of both chromosomes and nucleolar material occurs during anaphase possibly as a result of a 'pushing' effect by the spindle microtubules (fig. 119). At the end of anaphase and during early telophase the daughter nuclei become enclosed once again by a reconstituted envelope (fig. 120) and the nucleolar material recondenses. The spindle pole body becomes associated with the reconstituted nuclear envelope (fig. 121). The break down of the nuclear envelope during mitosis in *B. ranarum* is a feature which, for fungi, has so far been demonstrated only in Basidiomycotina (Section III, p. 199). It may prove to be atypical for other Zygomycotina.

[1] Since information on the internal ultrastructure and role of this organelle is at present lacking, and since *B. ranarum* is a 'non-flagellate' fungus, the term **spindle pole body** will be used for it (see footnote Section II, p. 93).

Additional reading

TANAKA, K. (1970). Mitosis in the fungus *Basidiobolus ranarum* as revealed by electron microscopy. *Protoplasma*, **70**, 423.

Fig. 115
An interphase nucleus of *Basidiobolus ranarum* Eidam showing the large nucleolus and typical electron-transparent 'nucleolar vacuole'. × 9,000.

Fig. 116
A prophase nucleus of *B. ranarum* in which the nucleolar material has become dispersed throughout the nucleoplasm. × 10,160.

Fig. 117
A nucleus at metaphase. The chromosomes lie on the metaphase plate (between arrows) and on either side is the nucleolar material (Nu). × 12,250.

Fig. 118
A spindle pole body (SPB) is lying 'side on' to the metaphase plate. × 18,570.

Fig. 119
An anaphase nucleus of *B. ranarum* showing separation of daughter nuclei. The nuclear envelope (Ne) has re-formed around part of the nucleolar material. The spindle microtubules (arrows) have moved apart leaving a central area apparently free of microtubules (open arrows). × 10,450.

Fig. 120
A telophase nucleus in which the nuclear envelope almost completely surrounds the condensing nucleolar material. Note microtubules which mark the remnants of the spindle (arrows) and the spindle pole body (SPB). × 13,200.

Fig. 121
A late-telophase nucleus. The characteristic 'interphase morphology' is almost restored. Note spindle pole body (SPB) associated with the nuclear envelope. × 12,050.

Figs. 115–21 *Glutaraldehyde–osmium tetroxide fixation. Micrographs by* DR K. GULL, Queen Elizabeth College, London.

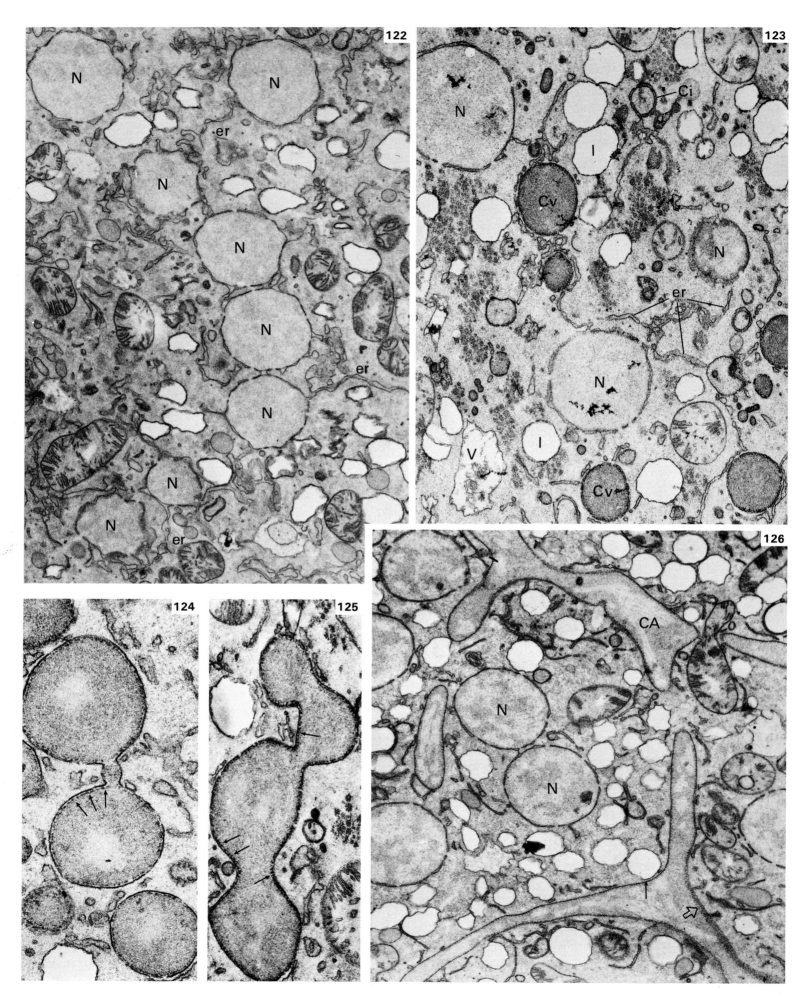

B. Reproductive Structures
Asexual reproduction
Sporangiospore production

In Zygomycotina, for example *Gilbertella persicaria* (Eddy) Hesseltine, asexual reproduction involves a **cleavage** of the cytoplasm of the normally spherical sporangium to produce numerous nucleate, non-motile sporangiospores (figs. 126, 127, 128). This process is essentially similar to that already described for Oömycetes (p. 31) and Chytritiomycetes (p. 35) in that all of the sporangial cytoplasm is utilized in the production of spores; however, there are certain differences in detail some of which are related to the fact that sporangiospores do not possess flagella.

Initially sporangiospore delimitation is **endogenous** since it begins within the cytoplasm by the association of membranes from the endoplasmic reticulum, nuclear envelope and subsequently from morphologically distinct **cleavage vesicles** (figs. 122–126). At the ultrastructural level it is possible to recognize certain distinct phases during the process of sporangiospore formation and these are summarized diagramatically in figure 129. During **early precleavage** extensive interconnections occur between membranes of endoplasmic reticulum and the nuclear envelope (fig. 122) resulting in a membrane complex which connects many nuclei together. These interconnections become less obvious during **late precleavage** (fig. 123) and as cleavage progresses (fig. 126). Early precleavage sporangia also possess small spherical vesicles around their edges which like the internuclear membrane continuities also tend to disappear after **mid-precleavage**. With the decrease of the above structures there is a corresponding increase in the numbers of larger cleavage vesicles which can be recognized by the presence of small, dense granules around the inner surface of the single, bounding membrane (figs. 123–126). These granules have been used as convenient morphological markers for studying the transformations and eventual fate of the cleavage vesicles up to the stage where mature sporangiospores are produced. Progressive fusion and extension of the cleavage vesicles during **early, mid-** and **late cleavage** phases (figs. 124–128) results in a complete division of the sporangial cytoplasm into nucleate portions (fig. 128). During mid-cleavage, the columella is also delimited from the rest of the sporangium by the **cleavage apparatus**.

The precise nature of the factors which control this process of **cytokinesis** is unknown (see also p. 31), but it has been suggested that in *G. persicaria* the continuity between nuclear envelopes and endoplasmic reticulum which surrounds cleavage vesicles indicates that nuclei play a role in the cleavage of protoplasm (see also Section II, Ascopore delimitation p. 157). It is thought that a selective association between cleavage vesicles and endoplasmic reticulum could lead to the development of vesicles surrounding nuclei and so influence the planes of cleavage and the boundaries of the new cells.

It follows that the process of cytoplasmic cleavage to form new cells (cytokinesis) necessitates the transformation of membranes of one type to that of another. For example, cleavage vesicle membrane becomes transformed during late cleavage to the plasma membrane (plasmalemma) of sporangiospore initials and eventually fuses with the plasma membrane of the sporangium itself. As a result of this development and transformation the originally intracellular cleavage vesicle granules become extracellular 'secretory' products.

Additional reading

BRACKER, C. E. (1968). The ultrastructure and development of sporangia in *Gilbertella persicaria*. Mycologia, **60**, 1016.

Fig. 122

Part of an early precleavage sporangium of *Gilbertella persicaria* (Eddy) Hesseltine. Eight nuclei (N) can be seen of which several have their envelopes interconnected by anastomosing sheets of endoplasmic reticulum (er). × 12,500.

Fig. 123

Part of a late precleavage sporangium showing the close association of nuclei (N), cleavage vesicles (Cv) and endoplasmic reticulum (er). Golgi cisternae (Ci) are also associated with endoplasmic reticulum. The presence of granules around the inside surface of the membrane facilitates the diagnosis of cleavage vesicles as compared with vacuoles (V) and lipids (I). × 16,000.

Figs. 124, 125

Stages in cleavage vesicle fusion. The dense granules lining the inside of the membrane are clearly visible (arrows). × 41,000 and × 32,000 respectively.

Fig. 126

Part of a mid-cleavage sporangium in which the cleavage apparatus (CA) formed by the fusion and extension of vesicles is dividing the sporangial cytoplasm into units containing one or more nuclei (N). Dense granules can be seen both in transverse (arrow) and tangential (open arrow) section. × 18,000.

Figs. 122–6 *Potassium permanganate fixation. From* BRACKER, C. E. (1968). *Mycologia*, **60**, 1016.

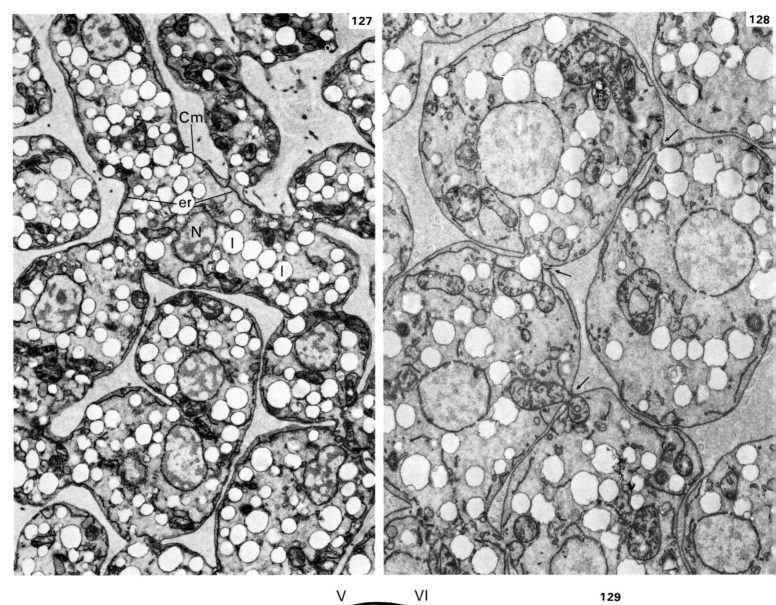

127 **128**

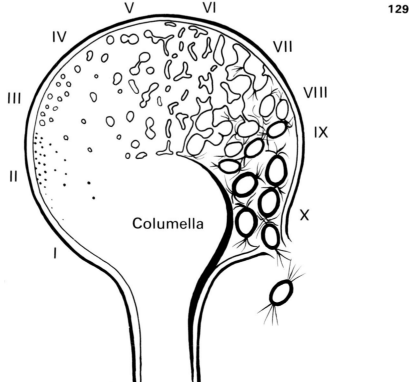

129

V VI VII

IV

III VIII

IX

II

Columella

I X

B. Reproductive Structures (cont.)
Asexual reproduction (cont.)
Sporangiospore production (cont.)

Fig. 127

Part of a sporangium nearing late cleavage. The cleavage apparatus is composed of a system of furrows which divide the sporangial protoplast into many units. Each unit contains one or more nuclei (N), many lipid globules (I) and endoplasmic reticulum (er) which lines the periphery of the individual units adjacent to the cleavage membrane (Cm). × 8,000.

Fig. 128

Part of a late cleavage sporangium. Cytoplasmic cleavage is almost complete with only narrow connections remaining between the spore initials (arrows). The granules lining the cleavage membrane now lie within interspore spaces (they will eventually become extracellular) and the cleavage membrane is now the plasma membrane of the spore initials. × 11,500.

Fig. 129

Gilbertella persicaria. Diagram summarizing the process of sporangiospore formation. Only the cleavage apparatus and walls are indicated. Spatial distribution of developmental stages is, for convenience, only in illustration, and does not represent spatial progression within a single sporangium. Early pre-cleavage (I, II), mid-pre-cleavage (III), late pre-cleavage (IV), early cleavage (V), mid-cleavage (VI), late cleavage (VII), early post cleavage (VIII), mid-post cleavage (IX), late post cleavage or spore maturity (X). The discontinuity in the sporangial wall near (X) represents rupturing of the sporangium as it splits into halves at maturity.

Figs. 127–9   (*127–8 Potassium permanganate fixation.*) From BRACKER, C. E. (1968). *Mycologia*, **60**, 1016.

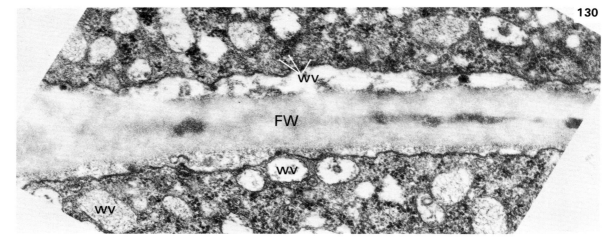

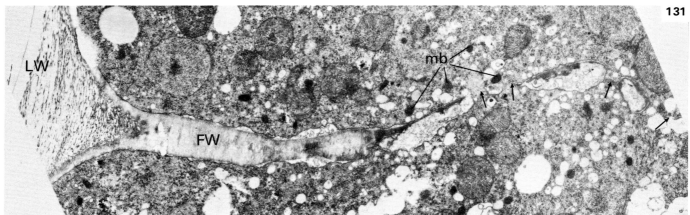

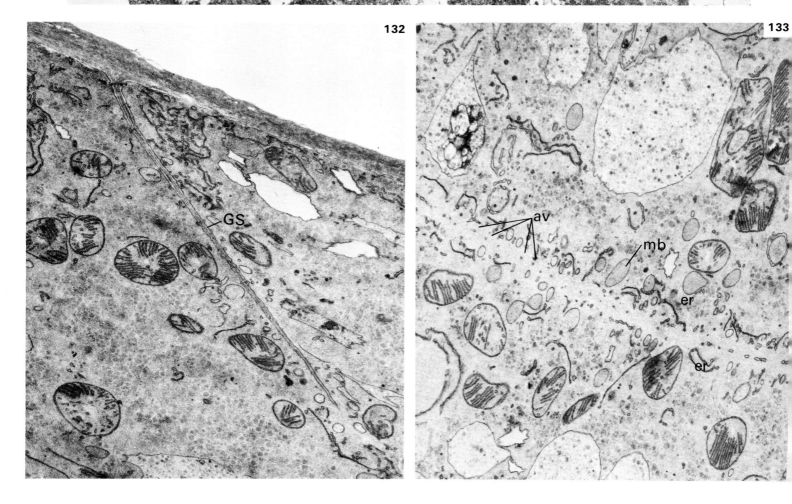

B. Reproductive Structures (cont.)
Sexual reproduction
Zygospore formation and structure

In *Rhizopus sexualis* (Smith) Callen zygospore formation is initiated by the apical conjugation of two aerial **zygophores**. After contact, the zygophores enlarge and change shape to form **progametangia**, the apical walls of which become adpressed to one another and fuse to form the **fusion wall**. Shortly after its formation the fusion wall, which is originally a flat plate of uniform thickness, becomes progressively broken down by a process of dissolution (fig. 130) which begins at the centre of the wall and spreads outwards (fig. 131). During the dissolution of the fusion wall there is a simultaneous synthesis of wall material further back from the apex of the two progametangia which results in the formation of cross walls or septa which delimit the **gametangia** (figs. 132, 133, 139 *c, d*). Microbodies and vesicles of various types are associated with both the breakdown of the fusion wall and the synthesis of the gametangial septum (figs. 130–133). The precise function of these vesicles is unknown but it has been suggested that some may transport hydrolytic enzymes to the fusion wall, which would be active in the subsequent dissolution, while others may transport degraded wall material away from the fusion wall possibly to be used during synthesis of the gametangial septum.

Gametangial septum formation involves the fusion of aligned vesicles at the advancing edge of the septum and it is thought that cytoplasmic streaming may play a part in initiating septum formation and in aligning the vesicles.

Deposition of **secondary wall** material leads to a thickening of the septum but since deposition occurs more rapidly and continues longer on the gametangial side of the septum the thickening is asymmetrical (fig. 134). A simultaneous wall thickening takes place along the lateral walls of the gametangia and after complete dissolution of the fusion wall the thick walled gametangia swell to form the **zygospore** (fig. 139 *d, e*). The gametangial septum is not strictly a complete septum but is perforated over its entire area by small tubular holes or **plasmodesmata** (figs. 134, 135). These plasmodesmata are often associated with large accumulations of endoplasmic reticulum on both sides of the septum and it is thought that soluble nutrients or other substances may pass from the **suspensors** into the developing zygospore via these perforations using endoplasmic reticulum membranes as a transport system.

Early stages in the development of the characteristic 'wart-like' ornamentations of the zygospore wall are recognized by the presence of numerous **lomasomes** along the inner edge of the wall together with an irregular deposition, within the wall, of electron-dense material (fig. 136). Progressive deposition of secondary wall material at specific sites around the expanding **primary wall** results in the formation of regions of dense material shaped like inverted flower pots (figs. 137, 138). Subsequently the outer layer(s) of the primary wall gelatinize and dissolve, thus exposing the raised, conical ornamentations (fig. 139 *e*).

Additional reading

HAWKER, L. E. and BECKETT, A. (1971). Fine structure and development of the zygospore of *Rhizopus sexualis* (Smith) Callen. *Phil. Trans. Roy. Soc. Lond.*, B, **263**, 71.

Fig. 130
Longitudinal section through parts of the apical region of two fused progametangia of *Rhizopus sexualis* (Smith) Callen. The fusion wall (FW), which is beginning to break down, is lined on both sides with numerous wall vesicles (wv) of various types. × 53,000.

Fig. 131
Longitudinal section showing a later stage in the dissolution of the fusion wall (FW). Note how the wall has broken down completely at various places towards the centre of the young gametangium (arrows), but still remains intact at the edge where it joins the lateral wall (LW). Numerous microbodies (mb) and various vesicles are concentrated in the region of maximum wall dissolution. × 10,000.

Fig. 132
Longitudinal section through part of a progametangium showing an early stage in formation of the gametangial septum (GS). × 15,500.

Fig. 133
Tangential longitudinal section through part of the gametangial septum showing the aligned vesicles (av) surrounded by microbodies (mb) and endoplasmic reticulum (er). × 15,500.

Figs. 130–3 *(130, 131 Glutaraldehyde/formaldehyde–osmium tetroxide fixation. 132, 133 Potassium permanganate fixation.) 130, 131 Micrographs by* A. BECKETT, *Bristol University. 132, 133 From* HAWKER, L. E. *and* BECKETT, A. *(1971). Phil. Trans. Roy. Soc. Lond.* B, **263**, 71.

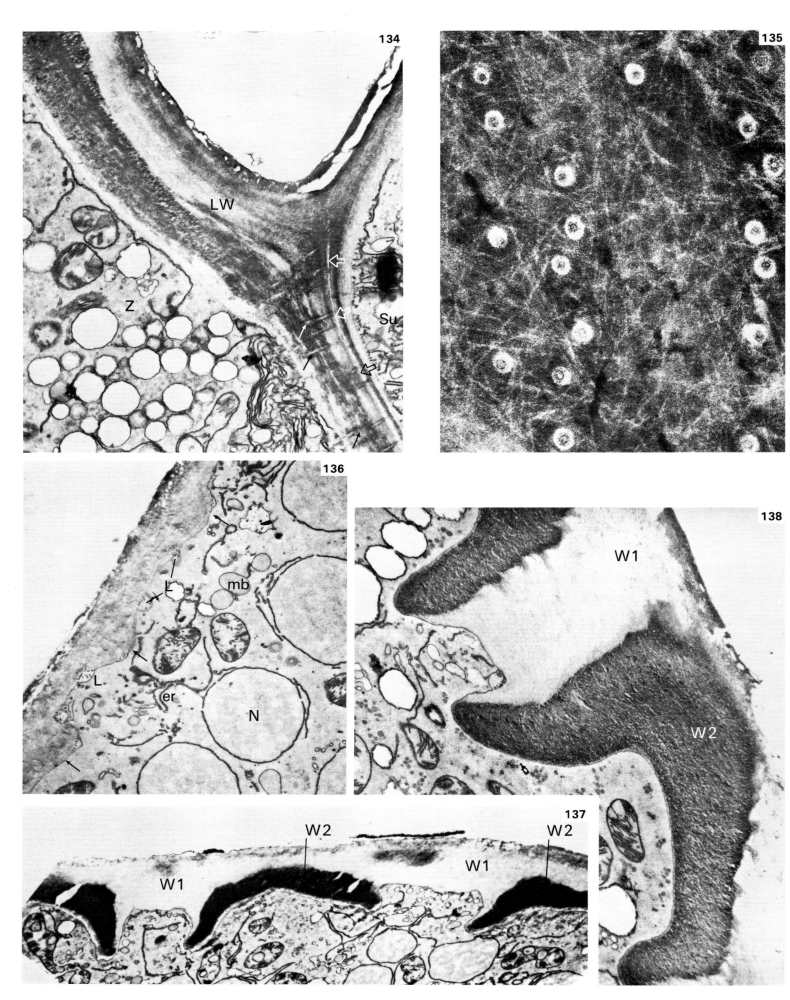

B. Reproductive Structures (cont.)
Sexual reproduction (cont.)
Zygospore formation and structure (cont.)

Fig. 134

Longitudinal section through part of the septum between the zygospore (Z) and suspensor (Su) at the point where it joins the lateral wall (LW). Note how secondary wall deposition has been greatest on the zygospore side of what was originally the middle, electron-transparent zone of the young wall (open arrows). Plasmodesmata can be seen in several places (arrows). × 15,000.

Fig. 135

Transverse section through part of the cross wall showing the tubular nature of the plasmodesmata which traverse the electron-dense secondary wall material. Note the random network of relatively electron-transparent fibrils within the wall. It is not known whether these are actual fibrils with a low affinity for stain or whether they may represent clear channels left in the matrix material by dissolution of fibrils during fixation. × 80,000.

Fig. 136

Part of a developing zygospore showing an early stage in deposition of secondary wall material at specific sites along the lateral wall (arrows). Lomasomes (L) are seen along the inner surface of the wall. Microbodies (mb), endoplasmic reticulum (er) and nuclei (N) are all found adjacent to the wall at this stage. × 10,800.

Figs. 137, 138

Parts of the zygospore lateral wall showing stages in the formation of the ornamentations of secondary wall material (W2) which are shaped like inverted flower pots and embedded within the inner layer(s) of primary wall (W1). × 6,500 and × 15,000 respectively.

Figs. 134–8 *Potassium permanganate fixation. 134, 135, 137 Micrographs by* A. BECKETT, Bristol University. *136, 138 From* HAWKER, L. E. and BECKETT, A. (1971). *Phil. Trans. Roy. Soc. Lond.,* B, **263**, 71.

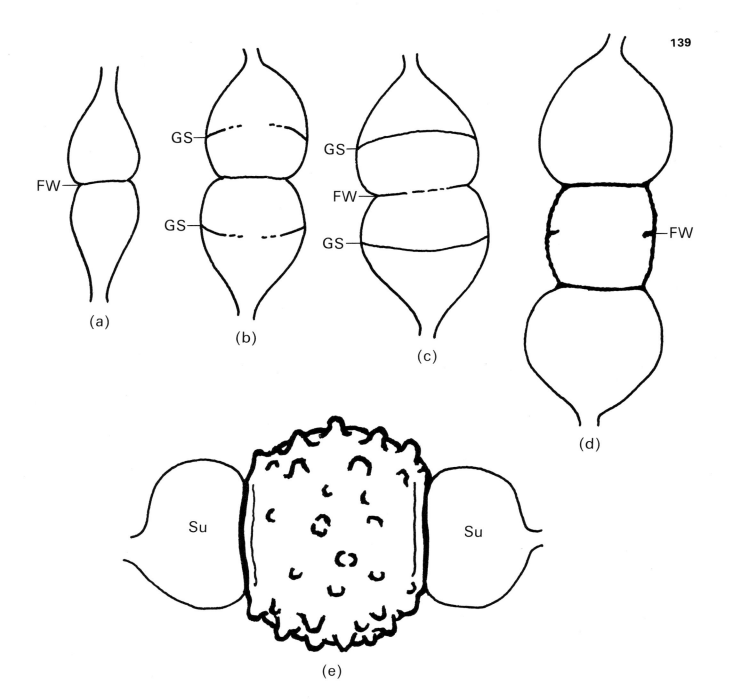

(a)

(b)

(c)

(d)

(e)

B. Reproductive Structures (cont.)
Sexual reproduction (cont.)
Zygospore formation and structure (cont.)

Fig. 139

Diagrammatic representation of zygospore formation in
R. sexualis. (a) Progametangia separated by the intact fusion
wall (FW). (b) Swelling of progametangia and initiation of the
gametangial septa (GS). (c) The gametangial septa (GS) are
complete and the fusion wall (FW) is partially broken down.
(d) The fusion wall (FW) has almost completely dissolved and
secondary wall deposition is beginning along the lateral and end
walls (gametangial septa) of the zygospore initial. (e) The
zygospore is now highly ornamented and readily distinguishable
from the smooth walled suspensors (Su).

Fig. 139 Modified from HAWKER, L. E. and BECKETT, A. (1971).
Phil. Trans. Roy. Soc. Lond. B, **263**, 71.

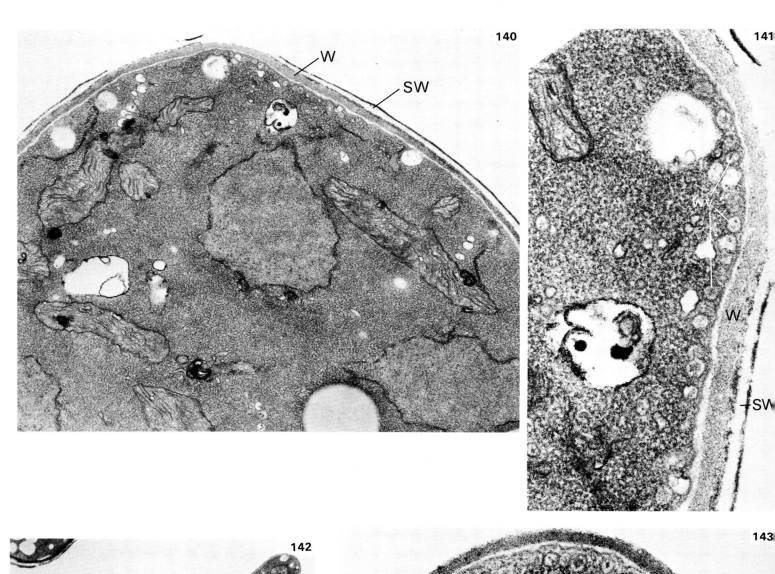

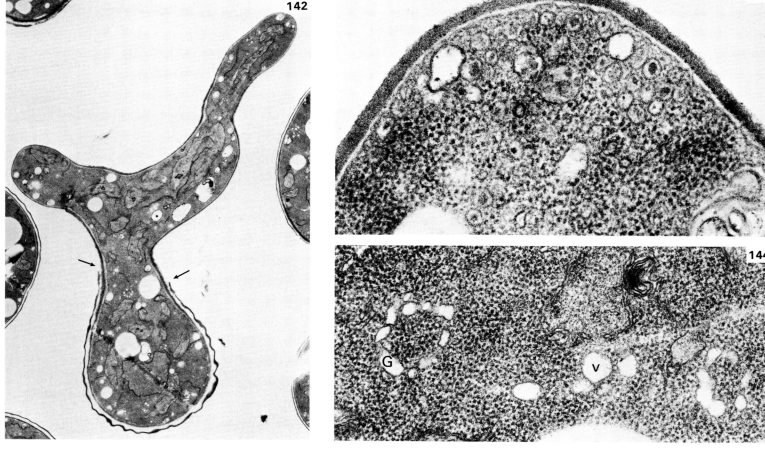

B. Reproductive Structures (cont.)
Sporangiospore germination

The formation of a germ tube during germination of sporangiospores of, for example, *Gilbertella persicaria* (Eddy) Hesseltine involves the accumulation of vesicles along the inside of the sporangiospore as has been described for Oömycete cyst germination (p. 45). Germination is initiated by a localized cell expansion, during which the outer wall of the sporangiospore is ruptured and a new inner wall forms the germ tube wall (fig. 140). It is at these sites of wall rupture that wall vesicles accumulate. Some vesicles have stainable contents, others do not (fig. 141). The newly formed inner sporangiospore wall grows out and around the germ tube as it emerges (fig. 142), and the growing germ tube tips contain a collection of vesicles (fig. 143) similar to those found in normal hyphal tips of established mycelia (fig. 110).

Golgi cisternae are found in the sub-apical regions of germ tubes associated with vesicles which do not possess stainable contents (fig. 144). Little is known of the chemical nature of the substance(s) within the vesicles but both types of vesicle are bounded by a membrane which is similar in density and dimension to the sporangiospore plasma membrane.

Additional reading

BRACKER, C. E. (1971). Cytoplasmic vesicles in germinating spores of *Gilbertella persicaria. Protoplasma*, **72**, 381.

Fig. 140

Part of a germinating sporangiospore of *Gilbertella persicaria* (Eddy) Hesseltine. The original sporangiospore wall (SW), which ruptures during germination, consists of an outer electron-dense layer and an electron-transparent layer. Prior to germ tube emergence, a new wall (W) forms which later extends out and around the germ tube. × 24,000.

Fig. 141

An enlargement of part of fig. 140 showing the accumulated wall vesicles (wv) beneath the plasma membrane in the region where the sporangiospore wall (SW) has ruptured. The contents of some vesicles stain in a similar way to the newly formed inner wall (W) which subsequently forms the germ tube wall. × 73,000.

Fig. 142

A germinated sporangiospore in which the wall of the branched germ tube is continuous with the newly formed inner wall of the sporangiospore. The ruptured edge of the original sporangiospore wall can be seen (arrows). × 7,000.

Fig. 143

Longitudinal section through a germ tube apex showing numerous wall vesicles. × 71,000.

Fig. 144

Golgi cisternae (G) and associated vesicles (v), the contents of which do not stain with uranyl acetate and lead citrate. × 64,000.

Figs. 140–4 *Glutaraldehyde/formaldehyde—osmium tetroxide fixation. From* BRACKER, C. E. (1971). *Protoplasma*, **72**, 381.

SECTION 2

Ascomycotina & Deuteromycotina

Introduction

Most Ascomycotina may be characterized by the possession of a regularly **septate vegetative mycelium** and by formation of endogenous **ascospores** during sexual reproduction. An apical body or **Spitzenkörper** is found in association with wall vesicles at the growing tips of hyphae. Golgi cisternae are possibly functional in the formation of wall vesicles during hyphal growth. Septa which form in somatic hyphae are perforate and lack the complex pore apparatus found in Basidiomycotina. (Section III, p. 193.) In **ascogenous** hyphae and at the base of asci septal pores are however associated with occluding structures of varying complexity.

Nuclear division involves the formation of an intranuclear spindle, the poles of which are associated with amorphous or plaque-like **spindle pole bodies** whose structure is unlike that of true centrioles. The latter are not known at any stage in the life cycles of higher fungi. Unlike Basidiomycotina and *Basidiobolus* (p. 61), the nuclear envelope remains intact until separation of daughter nuclei at telophase.

Asexual reproduction in most Ascomycotina is by formation of **conidia**, indeed, many form-genera of Deuteromycotina possess ascomycetous sexual states. For this reason conidium ontogeny in Deuteromycotina is considered within this section as a process of asexual reproduction.

The delimitation of ascospores by double membranes within an ascus (free cell formation) is unique to sexual reproduction in Ascomycotina and the subsequent discharge of these spores from the ascus involves in many cases structures of varying complexity, the development and ultrastructure of which are little known at present.

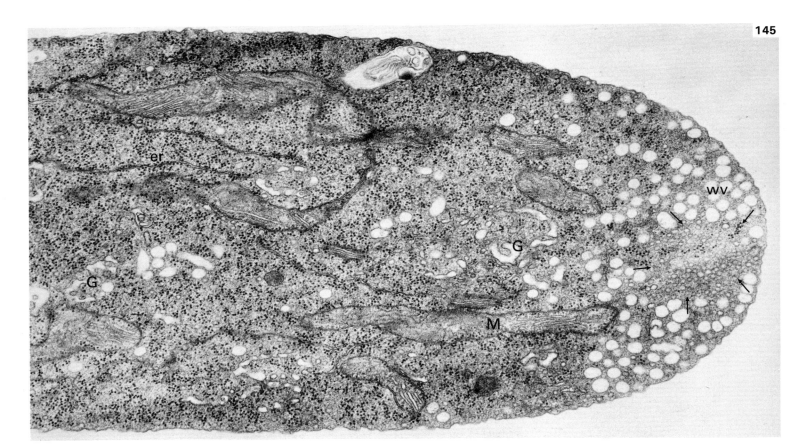

145

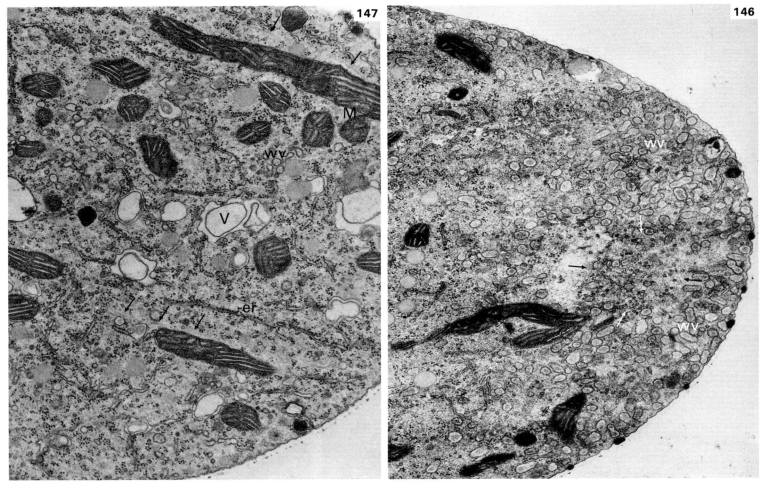

147

146

A. Vegetative Structures

Hyphae I. Cell wall synthesis

The vegetative phase of most Ascomycotina and Deutero-mycotina (with the exception of yeasts and yeast-like genera) is a system of regularly septate hyphal filaments. These filaments grow by a regular intersusception of wall material at the tip. In all Ascomycotina that have been critically studied, the hyphal apex is characterized by the possession of a system of wall vesicles together with a dense, amorphous body composed of small vesicles or granules which is termed the **Spitzenkörper** (figs. 145, 146). The Spitzenkörper has so far been demonstrated only in septate fungi (see also Section III, p. 185). The wall vesicles, which are thought to be formed by the Golgi cisternae, apparently fuse with the plasma membrane of the growing hypha, simultaneously contributing additional membrane to the plasmalemma and releasing their contents into the wall. It is thought that enzymes and/or wall precursor substances may be located within these vesicles and staining techniques have suggested that polysaccharides may be present in some cases (see also Section I, p. 5).

Recent work has suggested that a delicately controlled balance exists between synthetic and degradative processes which may occur at the hyphal apex. Osmotic shock, changes in pH and temperature can upset this balance and may lead to lysis of the cell wall. It is conceivable that wall vesicles could contain **hydrolytic enzymes** such as **glucanases** and **chitin hydrolases**, which, when released en masse at the plasma membrane—cell wall interface, could result in a break down rather than synthesis of wall components.

Additional reading

BARTNICKI-GARCIA, S. and LIPPMAN, E. (1972). The bursting tendency of hyphal tips of fungi: presumptive evidence for a delicate balance between wall synthesis and wall lysis in apical growth. *J. gen. Microbiol.*, **73**, 487.

GROVE, S. N. and BRACKER, C. E. (1970). Protoplasmic organization of hyphal tips among fungi: vesicles and Spitzenkörper. *J. Bact.*, **104**, 989.

Fig. 145

Median longitudinal section through the hyphal tip of *Aspergillus niger* van Teighem. The apical 1·0–2·0 μm is occupied by a mass of wall vesicles (wv) within which is a zone (delimited by arrows) of small vesicles and groups of ribosomes. This is the Spitzenkörper. Golgi cisternae (G), mitochondria (M) and rough endoplasmic reticulum (er) can be seen in sub-apical regions. The thin cell wall is not visible since it is electron transparent when fixed as in this case with glutaraldehyde/formaldehyde. × 38,000.

Fig. 146

Median longitudinal section through the apical region of a hypha of *Neurospora crassa* Shear and Dodge. The Spitzenkörper (arrowed) is situated immediately beneath the tip and is surrounded by a large number of apical wall vesicles (wv). × 23,000.

Fig. 147

Part of the sub-apical region of a hypha of *N. crassa*. Fewer wall vesicles (wv) occur in this zone but when present they are usually associated with Golgi cisternae. Numerous mitochondria (M) are found sub-apically usually elongated parallel to the hypha long axis and frequently associated with cytoplasmic microtubules (arrows). It is possible that mitochondria may migrate towards centres of metabolic activity along these microtubules. Early stages in vacuole formation (V) may also be seen. × 35,000.

Figs. 145–7 *Glutaraldehyde/formaldehyde—osmium tetroxide fixation. 145–6 From* GROVE, S. N. and BRACKER, C. E. (1970). *J. Bact.*, **104**, 989. *147 Micrograph by* DRS C. E. BRACKER *and* S. N. GROVE, Purdue University, Lafayette, Indiana.

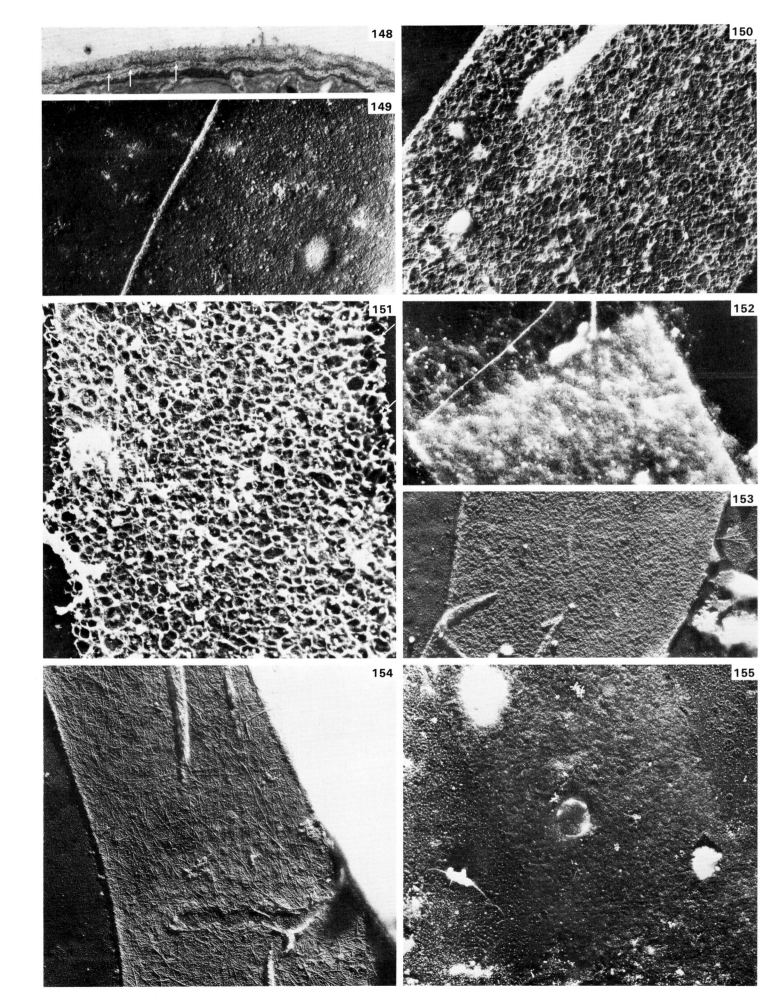

A. Vegetative Structures (cont.)

Hyphae II. Cell wall structure

Cell wall structure has been studied in *Neurospora crassa* Shear and Dodge by means of an enzymic-dissection technique similar to that described for *Phytophthora parasitica* (Section I, p. 9). *Neurospora* walls normally show a three-layered structure in section. Occasionally, indications of a fourth layer may be seen underneath the medium dense layer (fig. 148). *Neurospora* walls are known to contain **chitin**, **protein** and a **glucan** with β1,3 and β1,6 linkages. Enzymes employed here were **laminarinase**, **pronase** and **chitinase**. Control hyphae incubated in buffer alone show an amorphous, roughened appearance in shadow cast preparations (fig. 149). This appearance does not alter after single separate treatments with chitinase or pronase. Treatment with laminarinase, however, removes the outer amorphous wall region to reveal a network of coarse strands whose interstices are filled with amorphous material, giving a cratered appearance (fig. 150). Laminarinase treatment followed by pronase treatment results in a better definition of the reticulum strands, presumably by removing the amorphous material from the interstices

(fig. 151). In material treated only with laminarinase the reticulum appears in surface view to be coincident with the inner wall regions (see sides of hyphae in fig. 150), but, with subsequent pronase treatment, material is removed at the level of the reticulum and from beneath it giving rise to a clear region outside the innermost wall regions (fig. 151).

The effects of laminarinase/pronase treatment can also include dissolution of the reticulum (fig. 152), removal of which reveals an amorphous region of low electron density (fig. 153). The ultimate consequence of laminarinase/pronase treatment is the removal of this amorphous material to reveal microfibrils of chitinous dimensions (fig. 154). Subsequent treatment of this material with chitinase results in the degradation of these microfibrils (fig. 155). Figure 156 summarizes information obtained about hyphal wall structure in *Neurospora* from shadow cast preparations and also from sectioned material and compares this proposed wall structure with the wall layers that are visible in sections of untreated walls.

Additional reading

HUNSLEY, D. and BURNETT, J. H. (1970). The ultrastructural architecture of the walls of some hyphal fungi. *J. gen. Microbiol.*, **62**, 203.
MAHADEVAN, P. R. and TATUM, E. L. (1967). Localisation of structural polymers in the cell wall of *Neurospora crassa*. *J. Cell. Biol.*, **35**, 295.
MANOCHA, M. S. and COLVIN, J. R. (1967). Structure and composition of the cell wall of *Neurospora crassa*. *J. Bact.*, **94**, 202.
POTGIETER, H. J. and ALEXANDER, M. (1965). Polysaccharide components of *Neurospora crassa* hyphal walls. *Can. J. Microbiol.*, **11**, 122.

Fig. 148

Transverse section through part of the wall of an untreated cell of *Neurospora crassa* Shear and Dodge showing three major layers and an indication of a fourth thin layer (arrows). Glutaraldehyde–osmium tetroxide fixation. × 45,000.

Fig. 149

A shadowed preparation of cell walls of *N. crassa* which have been incubated in buffer only. Note the rough, amorphous surface. Pd/Au shadowed 40/60 Cot$^{1-3}$. × 10,000.

Fig. 150

A shadowed preparation of cell walls of *N. crassa* which have been treated with laminarinase. The amorphous, outer material has been removed leaving a network of coarse strands embedded in further amorphous material. Pd/Au shadowed 40/60 Cot$^{1-3}$. × 15,000.

Fig. 151

A shadowed preparation of cell walls of *N. crassa* which have been treated with laminarinase followed by pronase. The amorphous material surrounding the coarse strands has been

removed. Note the clear regions on each side of the hypha (arrows) formed as a result of the removal of material by pronase from beneath the reticulum. Pd/Au shadowed 40/60 Cot$^{1-3}$. × 15,000.

Figs. 152, 153, 154

Shadowed preparations of cell walls of *N. crassa* showing stages of dissolution of the reticulum and underlying amorphous material as a result of prolonged laminarinase/pronase treatment. Note microfibrils in fig. 154. Pd/Au shadowed 40/60 Cot$^{1-3}$. × 15,000.

Fig. 155

A shadowed preparation of cell walls of *N. crassa* which have been treated with laminarinase followed by pronase followed by chitinase. The microfibrils have been degraded by the chitinase. Pd/Au shadowed 40/60 Cot$^{1-3}$. × 15,000.

Figs. 148–55 *148 Micrograph by* DR D. HUNSLEY, Department of Agricultural Science, University of Oxford. *149–55 From* HUNSLEY, D. and BURNETT, J. H. (1970). *J. gen. Microbiol.*, **62**, 203.

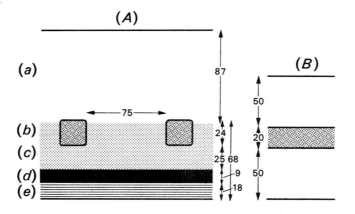

A. Vegetative Structures (cont.)

Hyphae II. Cell wall structure (cont.)

Fig. 156

(A) Reconstruction of a section through the wall of a hypha from a five day culture based on enzymic-dissection experiments. The numbers represent the mean thickness of the layers in nm.
(*a*) Outermost layer of amorphous glucan containing β1,3 and β1,6 linkages. (*b*) Sections through coarse strands of (?) glycoprotein reticulum. (*c*) Easily removable protein in which reticulum is embedded: there is thought to be an increasing concentration of protein from the outer part of this region inwards.

(*d*) A discrete layer of protein. (*e*) Innermost layer of chitin microfibrils possibly intermixed with protein.

(B) Layers visible in a section of untreated wall from a five day culture fixed in glutaraldehyde—osmium tetroxide; the middle layer is more electron-opaque than the other two. The numbers represent the mean thickness of the layers in nm.

Fig. 156 *From* HUNSLEY, D. and BURNETT, J. H. (1970). *J. gen. Microbiol.*, **62**, 203.

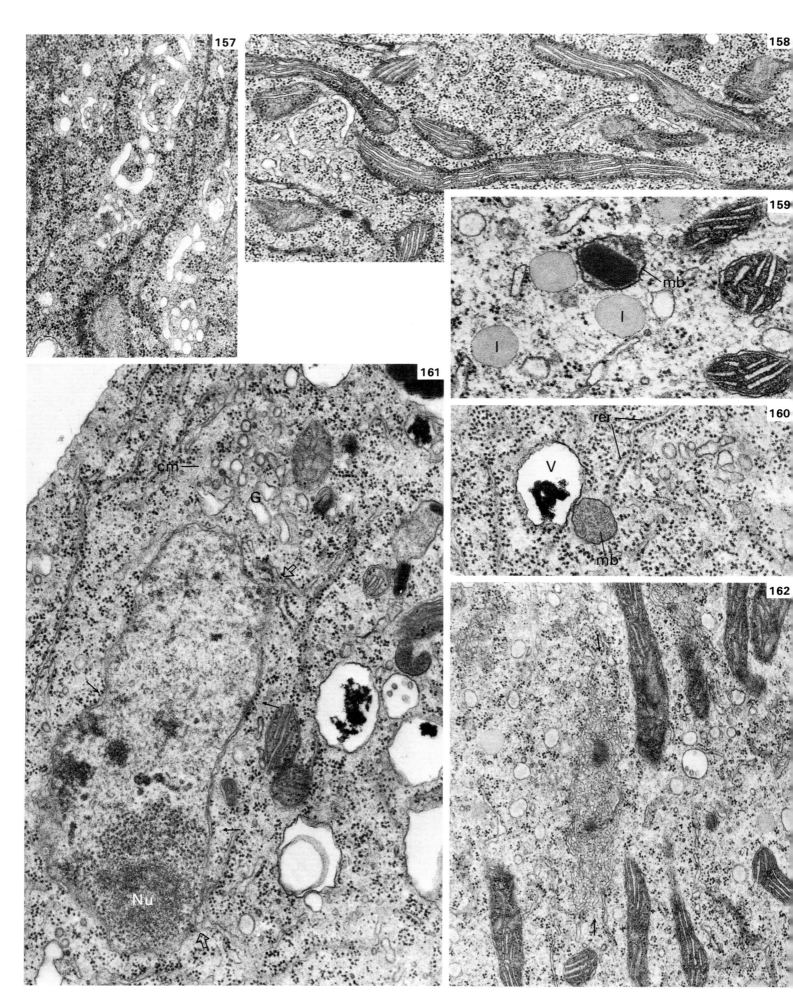

A. Vegetative Structures (cont.)

Hyphae III. Cell components

Mitochondria in most Ascomycotina and Deuteromycotina so far studied have been found to be highly elongated structures possessing numerous flat, plate-like cristae which are oriented parallel to the mitochondrion long axis. When fixed with glutaraldehyde/formaldehyde mixtures, the **perimitochondrial space** is usually electron-transparent and contrasts markedly with the typically electron-opaque mitochondrial **stroma** (figs. 158, 159, 162). As with all eukaryotic organisms, within the mitochondrial stroma are ribosomes which are smaller than the cytoplasmic ribosomes (figs. 158, 161, 162).

Nuclei in young, growing hyphae are often elongated parallel to the long axis of the hypha and contain a dense nucleolus at one end (fig. 161). The outer membrane of the nuclear envelope is typically studded with ribosomes along its outside surface (fig. 161 arrows) and this together with the fact that the nuclear envelope is continuous with the extensive membrane system of the endoplasmic reticulum has led to the belief that the nuclear envelope is a specialized region of endoplasmic reticulum.

Smooth membrane and vesicles occur throughout the cytoplasm in most Ascomycotina and Deuteromycotina. These may take the form of an irregular collection of vesicles and cisternae in *Neurospora crassa* (figs. 160, 161), or they may be arranged to form cisternal rings as in *Aspergillus niger* (figs. 157, 158). Both of these forms are considered to be **Golgi cisternae** and are components of the **Golgi apparatus**. Morphologically similar Golgi cisternae are found in Zygomycotina (Section I, figs. 114, 123, 144) and Basidiomycotina (Section III, figs. 349, 364, 365). **Membrane complexes** of variable form have been observed in a number of Ascomycotina both in vegetative hyphae as in *N. crassa* (fig. 162), and in ascospores as in *Ascodesmis sphaerospora* (fig. 323). Their

function(s) is unknown but they are continuous with the endoplasmic reticulum and probably represent a modified part of this membrane system.

Vacuoles are common in the subapical regions of hyphae (figs. 160, 161); they are surrounded by a single smooth membrane, the **tonoplast**, whose semipermeable properties enable vacuoles to function in many ways. They may contain hydrolytic enzymes and therefore function as **lysosomes**; they may contain food reserves; they may act as a depository for waste products formed by the cell, or they may act as a site for accumulation and intermixing of substances which may later be utilized during cell metabolism. Recent work suggests that there may be many kinds of vacuoles and that they may develop along diverse sub-cellular pathways. It is now clear that the early concept of the vacuole as being a simple 'dustbin' for the cell is inaccurate.

Microtubules (figs. 161, 162) are found throughout the cell cytoplasm with the exception of the extreme hyphal tip. Evidence is accumulating which suggests that there may be specific interactions between microtubules and other cell organelles such as nuclei, mitochondria and vesicles. One possible suggestion for such an association would be that microtubules act as sub-cellular 'tram lines' which guide and aid organelle motility (see also Section I, p. 13).

Other organelles which occur in hyphae are **microbodies**; small membrane-bound structures (0·1–2·0 μm diam.) with dense, finely granular contents and often crystalline inclusions (figs. 159, 160) and **lipid bodies**; (fig. 159), whose form varies considerably according to the method of fixation and probably also with the type of lipid present.

Fig. 157

Part of the sub-apical region of a hypha of *Aspergillus niger*. Smooth membrane Golgi cisternae are associated with vesicles similar to those found at the hyphal tip (see fig. 145). × 38,000.

Fig. 158

Part of a young hypha of *A. niger* showing the typical elongated mitochondria with longitudinally oriented cristae. × 35,000.

Fig. 159

Part of a young hypha of *Neurospora crassa* showing a microbody (mb) containing a crystalline inclusion, and three lipid bodies (l). × 60,000.

Fig. 160

Part of a young hypha of *N. crassa* showing rough endoplasmic reticulum (rer), smooth membranes and vesicles, a young vacuole (V) and a microbody (mb). × 50,000.

Fig. 161

Part of a young hypha of *N. crassa*. The elongated nucleus contains a nucleolus (Nu) composed of a granular ribosomal zone surrounding an amorphous electron-opaque zone. Dense patches in the nucleoplasm possibly represent chromatin. Ribosomes occur on the outside of the nuclear envelope (arrows) and the latter is continuous with rough endoplasmic reticulum (open arrows). A microtubule (cm) is associated with a collection of Golgi cisternae and vesicles (G). × 49,000.

Fig. 162

The sub-apical region of a hyphal tip of *N. crassa*, showing a smooth membrane complex (between arrows) which is continuous with rough endoplasmic reticulum and which contains two amorphous, granular regions at its centre. × 35,000.

Figs. 157–62 *Glutaraldehyde/formaldehyde–osmium tetroxide fixation. 157 From* GROVE, S. N. *and* BRACKER, C. E. (1970). *J. Bact.,* **104**, 989. *158–62 Micrographs by* DRS S. N. GROVE *and* C. E. BRACKER, *Purdue University, Lafayette, Indiana.*

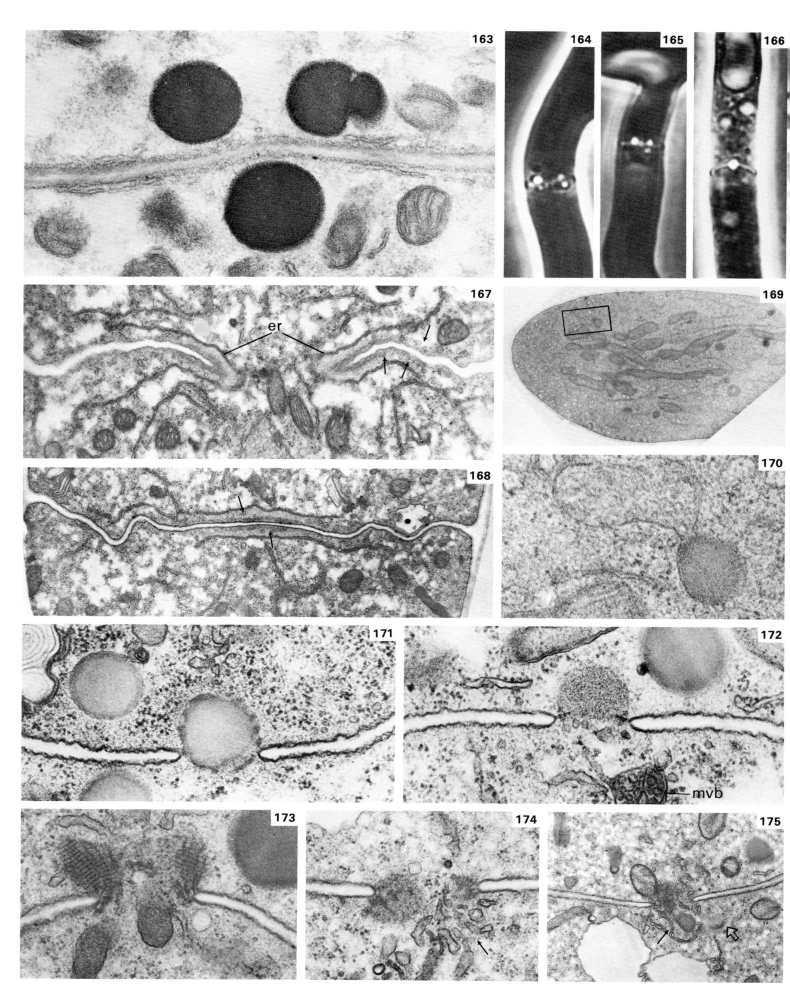

A. Vegetative Structures (cont.)

Septa and associated structures

Septa have been found within the vegetative structures of all Ascomycotina and Deuteromycotina. They are specialized, dynamic, taxonomically useful structures, capable of controlling or influencing translocation, sexuality, migration of organelles and senescence.

Septa in vegetative hyphae of Ascomycotina are typically perforate; frequently highly refractile spherical bodies known as **Woronin bodies** are found associated with pores. In *Ascodesmis sphaerospora* Obrist. the development of the septum and associated structures has been traced by fixation and embedment of mycelial mats and subsequent observation of thin sections cut from blocks taken at 1 mm intervals from the hyphal tips to a zone 7 mm behind the hyphal tips. In this fungus the process of septum formation occurs very rapidly. A cylinder of endoplasmic reticulum appears to be intimately involved in the process. A zone of vesicles appears adjacent to the forming septum and may be involved in wall deposition (figs. 167, 168). Woronin bodies are formed in the hyphal apices within single-membrane bounded sacs (figs. 169, 170), although they do not appear to associate with septal pores immediately. In older cells they are invariably found a short distance from the lumen of the pore (figs. 163–165). They are usually spherical or oblong and show a well-defined periodic substructure when sectioned in the appropriate plane (fig. 163). When a cell is injured, the Woronin bodies appear to coalesce and plug the pore (fig. 166). In cells of *Ascodesmis* 7–13 hours old signs of cell senescence and impending death are found. In such cells numerous anomalous profiles occur adjacent to the rim of the septal pore. These structures are thought to represent stages in the degradation of Woronin bodies. Within a given cell the process appears to take place serially, each of the several Woronin bodies in the cell moving into the septal pore one at a time and degenerating (figs. 171–175). Although the process is by no means synchronous, one encounters Woronin bodies with ever decreasing frequency as the age of the cells increases. In the oldest cells examined they are seldom seen.

Additional reading

BRACKER, C. E. (1967). Ultrastructure of Fungi. *A. Rev. Phytopathol.*, **5**, 343.
BRENNER, D. M. and CARROLL, G. C. (1968). Fine-structural correlates of growth in hyphae of *Ascodesmis sphaerospora. J. Bact.*, **95**, 658.

Fig. 163

Woronin bodies associated with an old cross wall (approx. 5 hours) of *Ascodesmis sphaerospora*. Note substructure evident in lower of the three Woronin bodies. ×52,000.

Figs. 164, 165

Phase contrast micrograph of an injured hyphal segment of showing septa and associated Woronin bodies in cells 4–9 hours old. ×1,700.

Fig. 166

Phase contrast micrograph of an injured hyphal segment of *A. sphaerospora* showing Woronin bodies plugging the pore. ×1,700.

Fig. 167

Part of a hypha of *A. sphaerospora* showing a late stage in cross-wall formation. Note endoplasmic reticulum (er) associated with the advancing rim of the septal pore and vesicles (arrows). ×20,000.

Fig. 168

Non-median section through a recently formed septum of *A. sphaerospora*. Note lateral extension of endoplasmic reticulum and vesicles (arrows) associated with the pore rim. ×17,200.

Fig. 169

Longitudinal section of a hyphal tip of *A. sphaerospora* showing Woronin body formation from a membrane bound sac (within rectangle). ×12,000.

Fig. 170

Enlargement of area within rectangle in fig. 169. ×90,000.

Figs. 171–5

Successive stages in Woronin body degradation in old, senescent hyphal cells of *A. sphaerospora*. Note in later stages proximity of multivesicular bodies (mvb), probable endoplasmic reticulum–lysosome complexes (arrows) and Golgi cisternae (open arrow) to the septal pore. Figs. 171–4. ×40,000. Fig. 175. ×15,000.

Figs. 163–75 *(163, 167–75 Glutaraldehyde–osmium tetroxide fixation.) 163–70, 172–4 From* BRENNER, D. M. and CARROLL, G. C. (1968). *J. Bact.* **95**, 685. *171 and 175 Micrographs by* DRS D. M. BRENNER *and* G. C. CARROLL, *University of Oregon, Eugene, Oregon.*

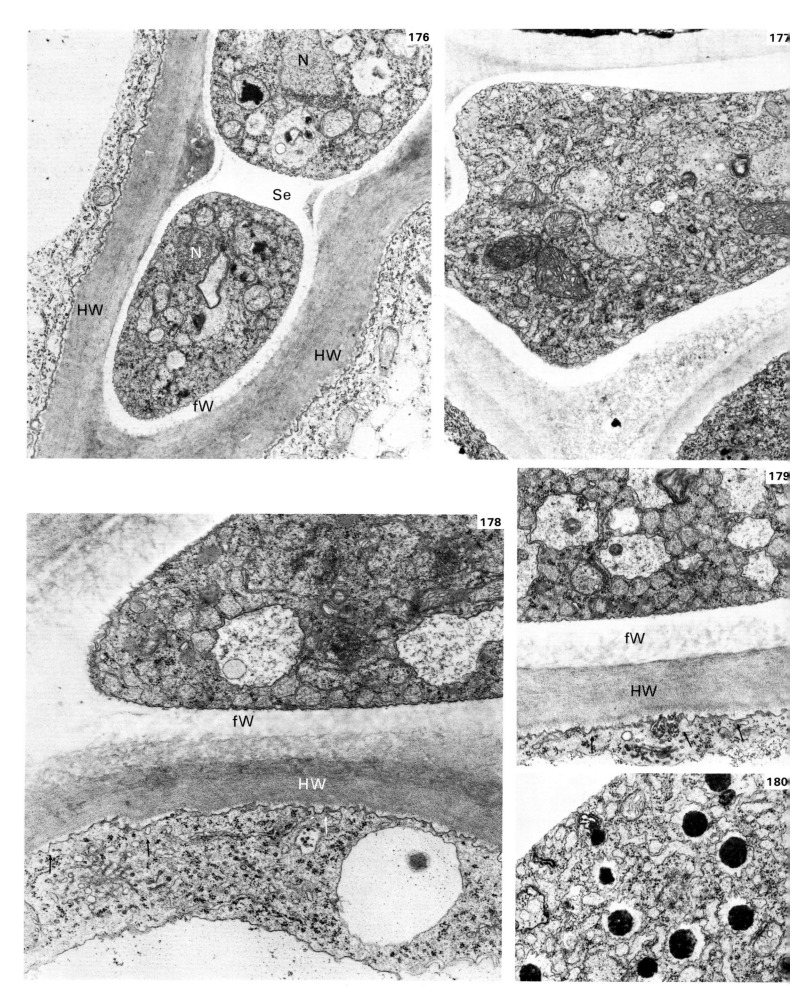

A. Vegetative Structures (cont.)

Intercellular hyphae

Some fungi are pathogens of higher plants and as such complete much of their life cycle within the **intercellular spaces** of leaves and stems. *Taphrina deformans* (Berkeley) Tulasne is an ascomycete fungus which causes a leaf curl disease of *Prunus amygdalus* and *P. persica*. The vegetative mycelium is made up of short, septate cells and ramifies between the cells of the host leaf although never becoming **intracellular**. The cell walls are electron-transparent when fixed with most of the commonly used fixatives. The cells are separated by wide cross walls (fig. 176) and it has not yet been shown whether these septa are complete or perforate. The intercellular hyphae within the **mesophyll** contain many cisternae, vesicles, nuclei and several large vacuoles (figs. 178–180) and in some cells large deposits of electron-opaque material may be seen within vacuoles which are surrounded by a typically crenulated membrane (fig. 180). **Sub-epidermal** hyphae possess similar organelles but generally have fewer vesicles and cisternae, the latter are less dilated in these cells and often connect with endoplasmic reticulum (fig. 177). Mycelial cells which have grown up between the epidermal cells to occupy the **sub-cuticular** zone, contain little or no vesicles and cisternae but do possess numerous mitochondria, one or more nuclei and densely packed cytoplasmic ribosomes (figs. 176).

Callose deposition, the typical defense reaction of the host in response to infection (see Section I, p. 21), is not found in host cells infected with *T. deformans*. There is however a marked stimulation of Golgi activity within the host which eventually results in an increase in cell division within the epidermis and mesophyll and finally hypertrophy of the leaf.

Figs. 176–180
Transverse sections through parts of a leaf of *Prunus amygdalus* infected with *Taphrina deformans* (Berkeley) Tulasne.

Fig. 176
Part of a sub-cuticular hypha of *T. deformans* showing the typical wide, electron-transparent septum (Se) and the dense cytoplasm containing numerous ribosomes, mitochondria and a few vesicles. A single nucleus (N) can be seen in each cell. Note the intimate association between the fungus cell wall (fW) and the walls of the two adjacent epidermal cells (HW). × 12,500.

Fig. 177
Part of a sub-epidermal hyphal cell in which there are numerous cisternae formed by dilation of endoplasmic reticulum. × 15,500.

Figs. 178, 179
Parts of intercellular hyphal cells from within the host mesophyll showing masses of more or less spherical vesicles, dilated cisternae and several large vacuoles. Note the close association between the fungus cell wall (fW) and host wall (HW) and the apparent fusion of Golgi vesicles with the host plasma membrane (arrows). Both figures × 23,000.

Fig. 180
Part of an intercellular hyphal cell which contains numerous cisternae and several vacuoles within which are deposits of electron-opaque substance. × 23,500.

Figs. 176–80 *Glutaraldehyde/formaldehyde–osmium tetroxide fixation. Micrographs by* MARY SYROP, Department of Botany, Bristol University.

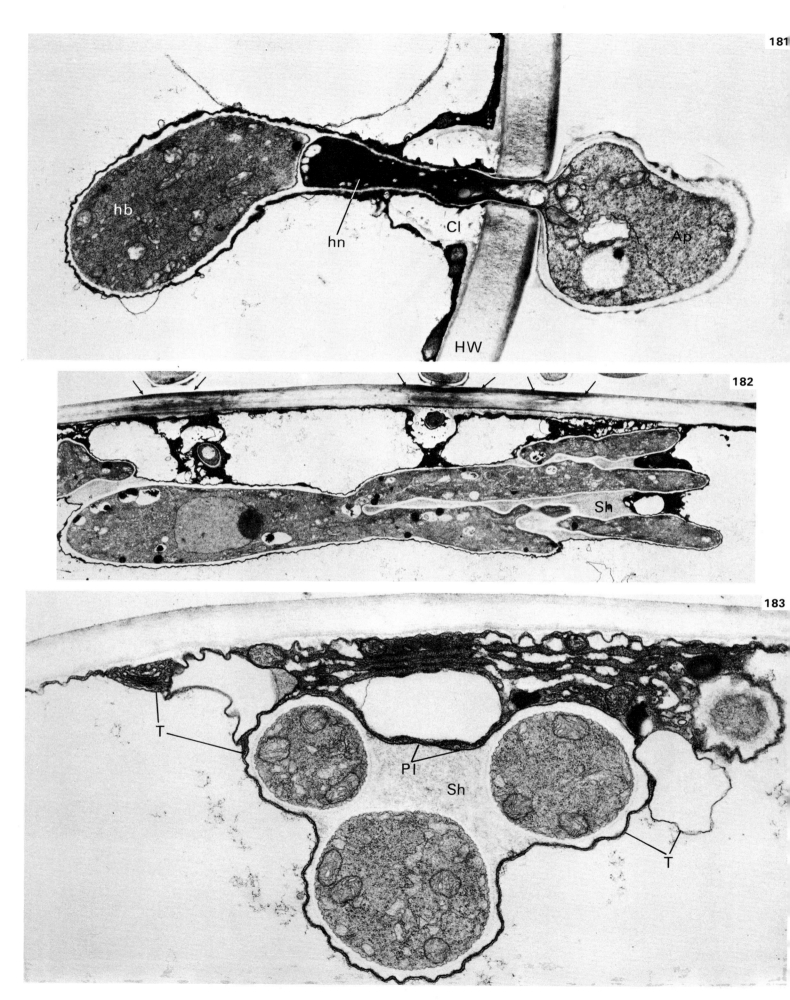

A. Vegetative Structures (cont.)

Haustoria

The 'powdery mildews' (Erysiphales) are highly specialized haustorium-forming parasitic Ascomycotina. They grow almost entirely superficially over the surfaces of the host plant, penetrating the host cells in numerous places to form **intracellular** haustoria. *Erysiphe graminis* de Candolle ex Merat, which grows on various cereals and grasses, produces on the epidermal surface an **appressorium** from which a hyphal filament grows and penetrates the cell wall (fig. 181). The distal end of this filament (the **haustorial neck**) swells to form the **haustorial body**. A septum (fig. 181), which in median section can be seen to be perforated, delimits the body from the neck region. The proximal end of the haustorial neck is surrounded by a collar-like deposit upon the host wall (fig. 181) (see also Section III, fig. 381). The chemical composition of this collar material is unknown, but structurally it is clearly different from the host cell wall.

Suitable longitudinal sections show the nucleate haustorial body to be multilobed (fig. 182). The entire haustorium is surrounded by a **'sheath'** and usually a thin layer of host cytoplasm. The sheath material is enclosed by the invaginated host plasma membrane (fig. 183) and so although the haustoria are intracellular they lie outside the host protoplast, separated from it by plasma membrane (see also Section III, p. 195).

Additional reading

BRACKER, C. E. (1968). Ultrastructure of the haustorial apparatus of *Erysiphe graminis* and its relationship to the epidermal cell of barley. *Phytopathology*, **58**, 12.

BRACKER, C. E. and LITTLEFIELD, L. J. (1973). Structural concepts of host-pathogen interfaces. Ch. III.3 **159**. In: *Fungal Pathogenicity and the Plant's response*. (R. J. W. Byrde and C. V. Cutting eds.), Academic Press, London.

Fig. 181

Transverse section through part of a wheat leaf epidermal cell infected with *Erysiphe graminis* de Candolle ex Merat. The appressorium (Ap) has produced a penetration peg which passes through the host cell wall (HW) to form the haustorial neck (hn) and haustorial body (hb). A ring of electron-transparent material associated with the inside of the host cell wall surrounds the proximal end of the neck to form the collar (Cl). Note the dense staining reaction with the host cell wall around the penetration site. × 12,000.

Fig. 182

Longitudinal section through a haustorium of *E. graminis* in an epidermal cell of wheat leaf. The body of the haustorium with a nucleus is connected to haustorial lobes. The section does not pass through a penetration point but the densely stained patches in the host cell wall (arrows) indicate regions where such sites exist beyond the plane of section. Note the sheath material (Sh) between the haustorial lobes. × 4,300.

Fig. 183

Transverse section of four lobes of a haustorium of *E. graminis* in an epidermal cell of a young wheat leaf. Three of the lobes are surrounded by a common sheath (Sh) which is bounded by an invaginated host plasma membrane (Pl). Between this and the host tonoplast (T) is a thin dense layer of host cytoplasm. × 21,000.

Figs. 181–3 *Glutaraldehyde–osmium tetroxide fixation. 181–2 From* BRACKER, C. E. and LITTLEFIELD, L. J. (1973) in *Fungal Pathogenicity and the Plant's Response*, Ch. III.3, pp. 159–317. Academic Press, London. *Micrograph by* DR C. E. BRACKER, Purdue University, Lafayette, Indiana.

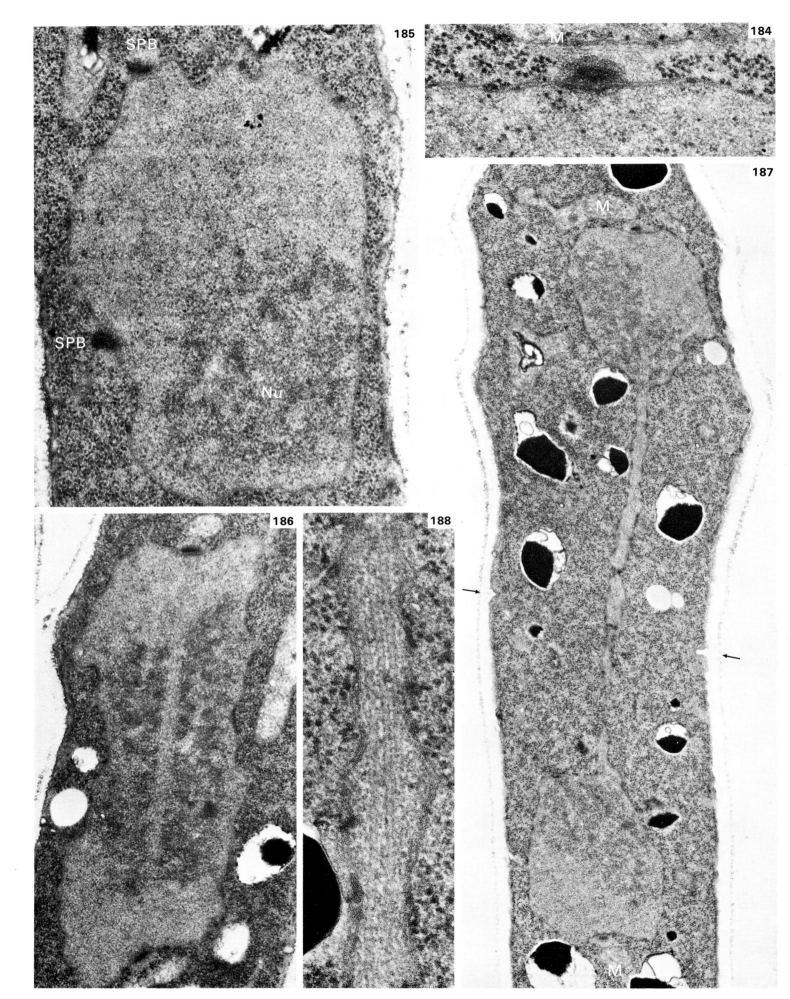

A. Vegetative Structures (cont.)

Mitosis

As with the flagellate fungi (Section I, p. 23), the mechanics of mitosis in higher fungi has been a controversial subject.

Two examples will be considered here.

In the ascomycetous fission yeast *Schizosaccharomyces pombe* Lindner the interphase nucleus is associated with a curved disk (*ca.* 220 nm in diameter) of electron-opaque material which lies in a ribosome-free zone close to the nuclear envelope on one side, and adjacent to a mitochondrion on the other side (fig. 184). This is the **spindle pole body**[1], a structure which undergoes replication prior to the onset of mitosis (see also Section II, p. 151).

During the early stages of mitosis, nuclei are characterized by the possession of a short, **intranuclear spindle** which joins two spindle pole bodies, each situated on the nuclear envelope a short distance apart (fig. 185). As mitosis proceeds the nucleus becomes elongated and rectangular in shape (fig. 186). The spindle now stretches from end to end across the centre of the nucleus. The component microtubules run parallel to each other in a close bundle between the two spindle pole bodies. The **nucleolus** persists and is stretched out in the interior of the nucleus (fig. 186). Further elongation results in the formation of a dumb-bell-shaped nucleus with two rounded ends joined by a long narrow channel containing the spindle microtubules (fig. 187). The widely separated spindle pole bodies remain at the poles of the microtubule bundle and each is still

[1] Throughout this atlas the term **spindle pole body** will be used to describe those structures of variable morphology found usually associated with the poles of the spindle during mitosis and meiosis, in **non-flagellate** fungi. True centrioles have not been found in these fungi. Spindle pole bodies are invariably associated with microtubules at least during part of the cell and nuclear cycle and as such may also play a role in nuclear migration and ascospore delimitation (see Section II, p. 157). Numerous terms for this structure exist in the literature and the reader is referred to the additional reading given below where this terminology is clarified. In addition one other synonym may be added here: Spindle pole body = Archontosome (Beckett and Crawford 1970. See additional reading Section II, p. 157).

associated with the nuclear envelope and a mitochondrion (fig. 187). Little is known at present of the significance of this association with mitochondria.

At no stage of mitosis are recognizable **chromosomes** evident in electron micrographs. However, by comparison with chromosome configurations in stained light microscope preparations of dividing nuclei, an impression has been gained that the chromosomes move apart in two clusters towards the poles of the spindle because they are in fact attached to the spindle pole bodies. Alignment of the spindle pole bodies may in turn be brought about by differential membrane growth in the region of the nuclear envelope which lies between them, and their subsequent separation may be the result of a general growth and expansion of the nuclear envelope. It follows that in order for such a process to operate efficiently the nuclear envelope must remain intact throughout division until the final separation of the daughter nuclei occurs. However, an alternative theory suggests that the elongation of spindle microtubules provides the necessary forces during separation of daughter nuclei at telophase (see below).

Mitosis in the vegetative mycelium of the Deuteromycete *Fusarium oxysporum* Schlect. is somewhat more closely allied to the so called 'classical' mitotic pattern in that distinct chromosomes can be recognized which are connected to specific microtubules of the spindle (**chromosomal microtubules**) at the kinetochores (figs. 189, 190). Daughter chromosomes separate during anaphase across a microtubular spindle each pole of which is associated with an amorphous granular **spindle pole body**. The spindle pole body is often located within a 'pocket-like' indentation of the nuclear envelope.

During this anaphase separation of chromosomes the chromosomal microtubules become shorter so that the kinetochores and their attached chromosomes migrate towards one or other of the spindle pole bodies. Chromo-

Fig. 184

Longitudinal section through part of an interphase nucleus of *Schizosaccharomyces pombe* Lindner showing the single spindle pole body as a dense, curved disk with a bulge towards its centre. It is associated with the nuclear envelope on the lower side and a mitochondrion (M) on the upper side. × 80,500.

Fig. 185

Longitudinal section through a nucleus of *S. pombe* at an early stage of mitosis. The dense granular nucleolus (Nu) lies at one end of the nucleus and the chromatin (as determined by light microscopy of stained preparations) occupies the comparatively electron-transparent area. An intranuclear spindle lies to the left side of the nucleus and runs between the two spindle pole bodies (SPB). × 40,000.

Fig. 186

A later stage of division. The nucleus is elongated into a rectangular shape, the nucleolus is stretched out in the central

portion of the nucleus and the spindle lies along the long axis of the nucleus. × 28,000.

Fig. 187

Longitudinal section of a nucleus at a still later stage in mitosis. The elongate dumb-bell-shaped nucleus stretches almost the complete length of the cell. The nucleolar material has become distributed to the daughter dumb-bell ends which remain joined by a long narrow channel containing the spindle microtubules. Note the spindle pole body and adjacent mitochondrion (M) at each end of the nucleus, and the septum initials midway between the daughter dumb-bells (arrows). × 22,000.

Fig. 188

Part of the narrow elongated zone of the nucleus shown in fig. 187, showing spindle microtubules enclosed by the nuclear envelope. × 107,500.

Figs. 184–8 *Glutaraldehyde/formaldehyde–osmium tetroxide fixation. From* McCULLY, E. K. *and* ROBINOW, C. F. (1971). *J. Cell Sci.*, **9**, 475.

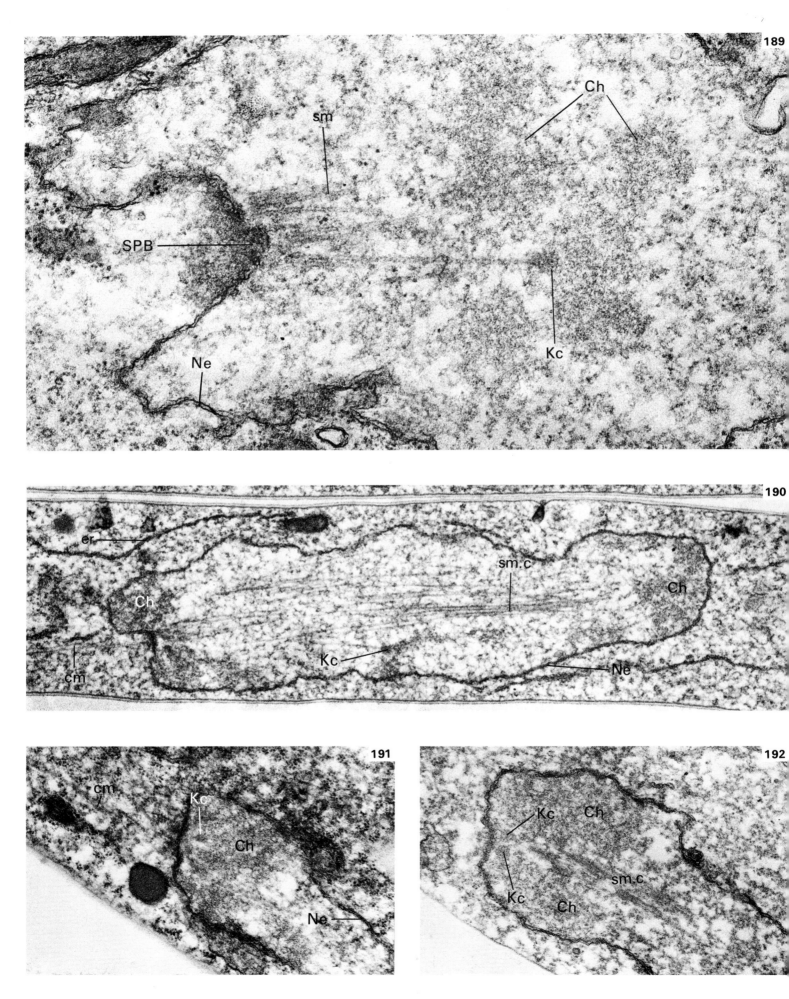

A. Vegetative Structures (cont.)

Mitosis (cont.)

somes in *F. oxysporum* are not attached at the nuclear envelope and the latter is not apparently involved in chromosome movement (cf. *Schizosaccharomyces pombe*). During telophase the pole to pole **continuous micro-tubules** of the spindle elongate and in so doing possibly cause the incipient daughter nuclei to move apart. The nuclear envelope finally 'pinches off' around these daughter nuclei. Cytoplasmic microtubules are associated with the spindle pole body during anaphase and telophase (figs.

190, 191, 193). Mitosis in *F. oxysporum* does however differ from the 'classical' pattern in the following respects: the nuclear envelope is **persistent** throughout the process (figs. 189–192); the chromosomes do not form a metaphase plate; there is an asynchronous chromatid disjunction during anaphase (fig. 190); and there is a possible connection between the chromosomes and a spindle pole body at interphase.

Additional reading

AIST, J. R. and WILLIAMS, P. H. (1972). Ultrastructure and time course of mitosis in the fungus *Fusarium oxysporum. J. Cell Biol.*, **55**, 368.
MCCULLY, E. K. and ROBINOW, C. F. (1971). Mitosis in the fission yeast *Schizosaccharomyces pombe*: a comparative study with light and electron microscopy. *J. Cell Sci.*, **9**, 475.

Fig. 189

Longitudinal section through part of a nucleus of *Fusarium oxysporum* Schlect. at metaphase of mitosis. Chromosomes (Ch), are associated with the spindle pole body (SPB) by means of spindle microtubules (sm). The spindle pole body is located within a 'pocket' of the nuclear envelope (Ne). One of the spindle microtubules is attached to a chromosome at a distinct kinetochore (Kc). × 1,10,000.

Fig. 190

A late anaphase nucleus of *F. oxysporum*. The daughter chromo-somes (Ch) have migrated to the spindle poles and numerous continuous spindle microtubules (sm.c) can be seen along the length of the nucleus. A lagging chromosome can be seen on the lower side of the nucleus together with its kinetochore (Kc).

Cytoplasmic microtubules (cm) can be seen associated with a spindle pole body. The nuclear envelope (Ne) is still intact. × 24,400.

Figs. 191, 192

Two of an incomplete series of sections through one end of a mid-telophase nucleus of *F. oxysporum*. The chromosomes (Ch) are densely packed and enclosed by the still intact nuclear envelope (Ne). Kinetochores (Kc), located at the spindle poles, are seen in side view (fig. 191) and in end view (fig. 192). A bundle of continuous spindle microtubules (sm.c) is seen in oblique section (fig. 192). Cytoplasmic microtubules (cm) are evident in fig. 191. Both figures × 34,400.

Figs. 189–92 Glutaraldehyde–osmium tetroxide fixation. From AIST, J. R. and WILLIAMS, P. H. (1972). *J. Cell Biol.*, **55**, 368.

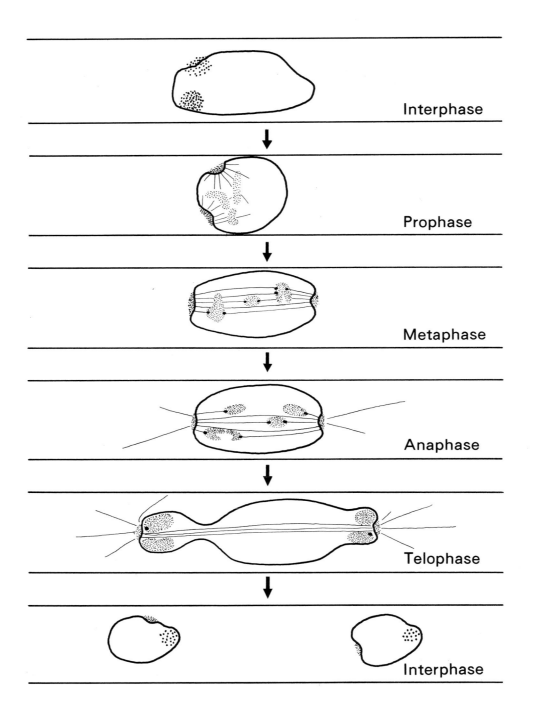

Interphase

Prophase

Metaphase

Anaphase

Telophase

Interphase

A. Vegetative Structures (cont.)

Mitosis (cont.)

Fig. 193

A semi-diagrammatic summary of mitosis in *F. oxysporum* based on light and electron microscope observations.

Fig. 193 *From* AIST, J. R. and WILLIAMS, P. H. (1972). *J. Cell. Biol.*, **55**, 368.

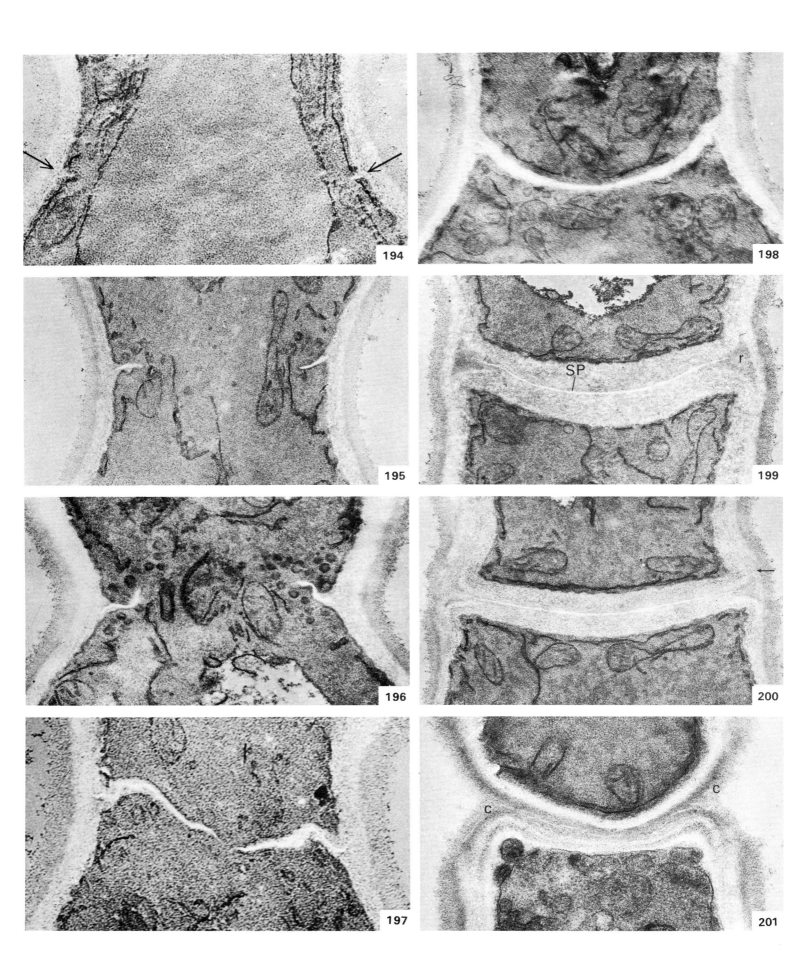

B. Reproductive Structures
Asexual reproduction I (budding)

Buds in the ascomycetous yeast *Saccharomycodes ludwigii* Hansen originate in one of two ways. They may arise as a 'blow out' or extension of the whole wall of the parent cell at its **distal end** (i.e. opposite end to the parent birth scar). In this case the bud is termed a **primary bud** (fig. 202, cell [4] arising from cell [2]). The wall at this stage is two layered and is covered by a granular capsule. A septum forms between bud and parent cell by the centripetal growth of a newly formed **ring** or **annulus** (figs. 194–197). The plasma membrane is continuous around the ingrowing septum (fig. 203). Closure of the annulus forms a complete disk-like cross wall, the **septal-plate** (fig. 198). The outer margin of the septal-plate is embedded in an electron-opaque ring, which is wedge-shaped in cross section (figs. 199–206). Separation of the bud from the parent is initiated by the rupture of the wall layer outside this ring (fig. 200). The cells part by the separation of the two halves of the septum along the line of the septal plate (fig. 201), which itself remains adhering to the wall of the parent cell

(fig. 204). Overlying the newly formed birth and bud scars of the daughter and parent cell respectively is the layer of electron-opaque, granular material continuous with both the remains of the ring and the capsular layer (figs. 201, 204).

The next bud arises at the proximal end of the parent cell by the extension of the half septum which forms the birth scar of the parent (fig. 202, cell [3] arising from cell [1]). The ruptured wall layers around the rim of the birth scar can be seen beneath the point where the bud arises (fig. 202). These buds and all subsequent ones from the one parent are termed **secondary buds**. Older cells may show multiple scars which signify repeated budding (fig. 205).

An interesting comparison may be made between the formation of the primary bud and its successors at the distal apex in *S. ludwigii* and the process of percurrent conidium formation in certain Deuteromycotina such as *Scopulariopsis brevicaulis* (Sacc.) Bain (see Section II, p. 123).

Additional reading

GAY. J. L. and MARTIN, M. (1971). An electron microscopic study of bud development in *Saccharomycodes ludwigii* and *Saccharomyces cerevisiae*. Arch. Mikrobiol., **78**, 145.

Figs. 194–8
Sections through part of a budding cell of *Saccharomycodes ludwigii* Hansen showing the sequence of development of the septum. An annulus forms by an electron-transparent ingrowth of the inner surface of the wall (fig. 194, arrows). This grows centripetally and becomes coated with other wall material finally forming a complete septum (fig. 198). × 40,000; × 32,000; × 32,000; × 46,000; × 34,000 respectively.

Fig. 199
A mature septum of *S. ludwigii* showing the septal plate (SP) terminating in the surrounding ring (r), which is wedge-shaped in cross section. × 15,500.

Fig. 200
A mature septum at the stage when the wall layer outside the ring has ruptured (arrow), cf. intact wall layer in fig. 199. × 14,500.

Fig. 201
Section through part of a budding cell of *S. ludwigii* in which the primary bud (upper cell) is separating from the parent by splitting of the septum along the septal plate. Note capsular material (c) covering the birth scar on the bud and the bud scar on the parent. × 21,000.

Figs. 194–201 *Potassium permanganate fixation. From* GAY, J. L. and MARTIN, M. (1971). *Arch. Mikrobiol.*, **78**, 145.

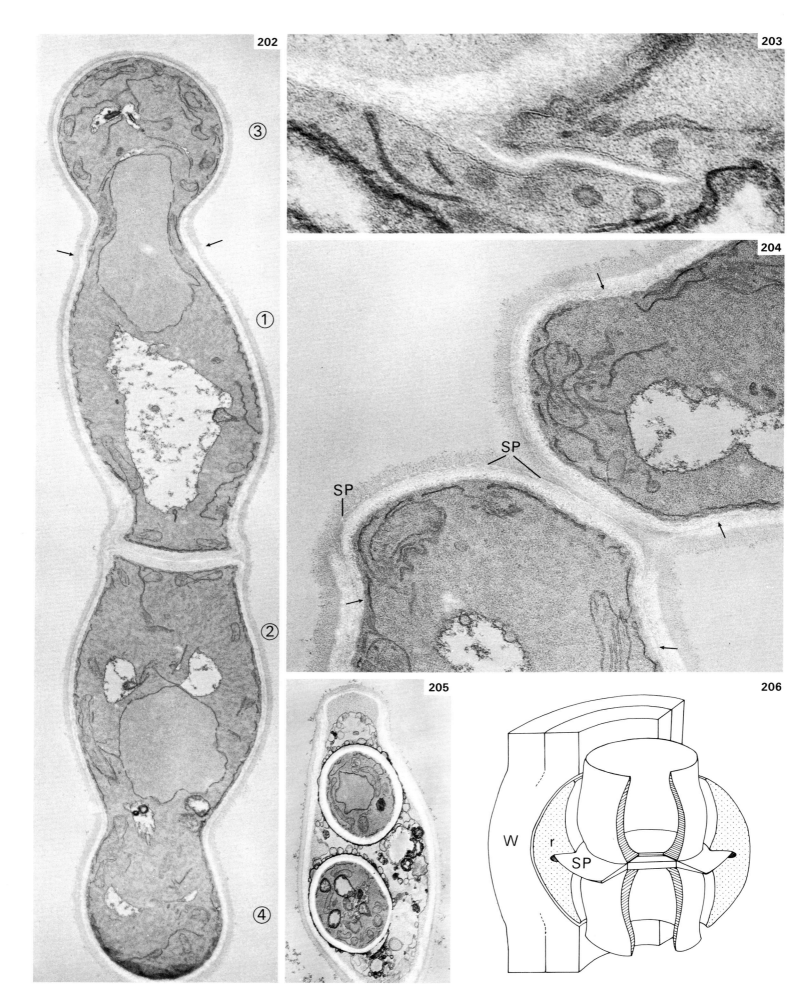

B. Reproductive Structures (cont.)
Asexual reproduction I (cont.)

Fig. 202

Longitudinal section through two cells of *S. ludwigii* showing two
types of bud development. The parent cell [1] has a birth scar
(arrowed) and cell [2] is its first daughter (not yet separated).
Cells [3] and [4] are the secondary bud of [1] and the primary
bud of [2] respectively. ×11,500.

Fig. 203

Part of an ingrowing cross wall showing the plasma membrane
continuous around the edge and several vesicles which may be
involved in wall deposition. ×102,000.

Fig. 204

Section of a primary bud (upper cell) separating from its parent.
The insertion of the 'half-septum' into the walls of the bud and
parent can be seen (arrows), and the septal plate (SP) is
attached to the bud scar on the parent. ×36,000.

Fig. 205

A maturing ascus of *S. ludwigii* showing multiple scars at the
upper end of the cell. ×10,000.

Fig. 206

Diagram of a sagittal section of a complete septum of *S. ludwigii*
showing wall structure immediately prior to cell separation:
(W) = cell wall; (r) = ring; (SP) = septal plate. Not drawn to
scale.

Figs. 202–6   (*202–5 Potassium permanganate fixation.*)
202–4, 206 From GAY, J. L. *and* MARTIN, M. (1971). *Arch. Mikrobiol.,* **78**, 145. *205 Micrograph by* DR J. L. GAY *and* M. MARTIN,
Imperial College, London.

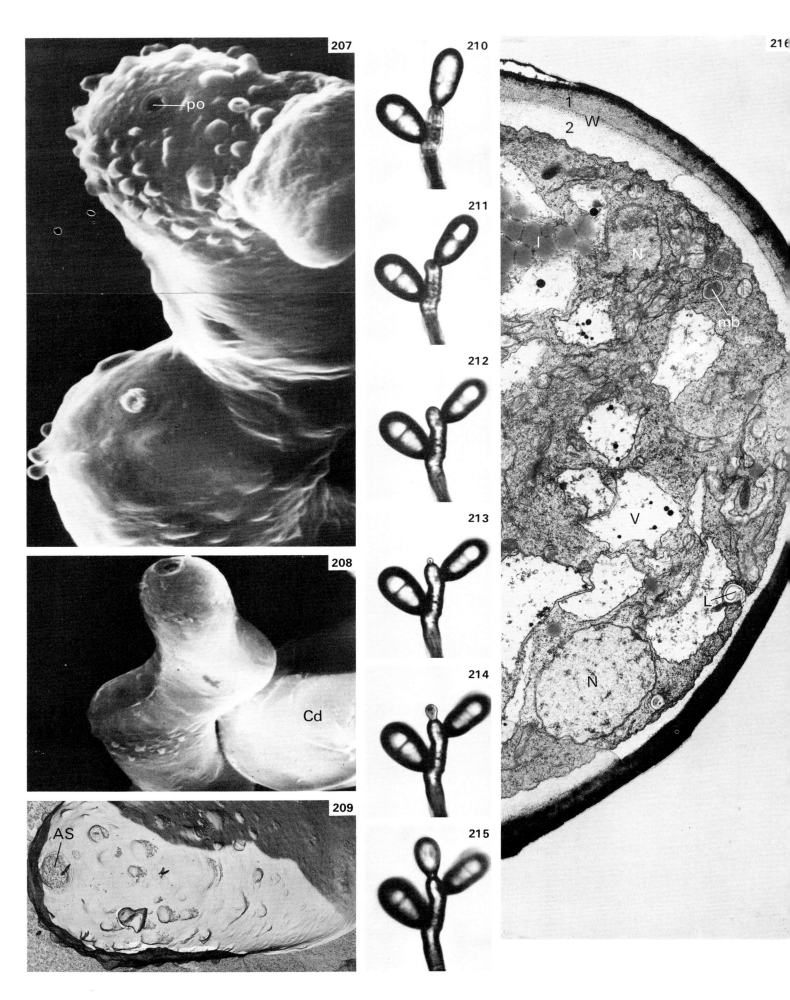

207

po

208

Cd

209

AS

210

211

212

213

214

215

216

1
2 W
l
N
mb
V
L
N

B. Reproductive Structures (cont.)

Asexual reproduction II (conidium production by Deuteromycotina)

A conidium may be defined as a specialized, non-motile, asexual propagule, usually deciduous and which is not formed by cytoplasmic cleavage (cf. sporangiospores, Section I, p. 63), or by free cell formation (cf. ascospores, Section II, p. 157). They are produced by a wide range of fungi representing all subdivisions of the Eumycota. Critical studies of conidium development at the ultra-structural level are at present few in number and most of these deal with but a few of the recognized ontogenetic patterns. The examples presented here are not therefore intended to be comprehensive, but will merely serve to illustrate some of the developmental processes involved and to exemplify some of the features currently used for taxonomic purposes. For details of the types of conidium ontogeny found in Deuteromycotina and of the relevant terminology, see additional reading below.

The fungus *Drechslera sorokiniana* (Sacc.) Subram. and Jain forms conidia as 'blow outs' through a narrow **channel** in the outer wall layers of the **conidiogenous cell**. It is thought that these channels are formed by **autolysis** of the wall. This means therefore that the outer wall layers of the conidiogenous cell are not involved in the formation of the **conidium wall**, the latter arising as an extension of the **inner wall** layer of the **conidiogenous cell**.

Conidia are formed in an acropetal sequence, each from a new fertile apex of the conidiogenous cell. The conidiogenous cell grows **sympodially** as a result of successive **proliferations** which occur from alternating sides just below each differentiated conidium (figs. 207, 208). Such a growth sequence is well illustrated by time-lapse micrography and seen in *Curvularia inaequalis* (Shear) Boedijn (figs. 210–215).

Scars are formed at fertile loci where conidia have been released (fig. 207) and at the base of the conidium (fig. 209). As conidia mature a new wall layer (**secondary wall layer**) is laid down by **apposition** within the original, **primary layer** (fig. 216).

Additional reading

COLE, G. T. (1973). Ultrastructure of conidiogenesis in *Drechslera sorokiniana. Can. J. Bot.,* 51, 629.
ELLIS, M. B. (1971). Dematiaceous Hyphomycetes. Commonwealth Mycological Institute, Kew, Surrey, England. pp. 1–608.
KENDRICK, B. (1971). *Taxonomy of Fungi Imperfecti.* University of Toronto Press, pp. 1–309.

Fig. 207

Scanning electron micrograph of a sympodially proliferated conidiogenous cell of *Drechslera sorokiniana* (Sacc.) Subram. and Jain. The abscission scars of two previously formed conidia are shown. The lower scar is occluded by thickened wall material and the upper, younger scar still shows a pore (po) through the conidiogenous cell wall. A newly proliferated region of the conidiogenous cell is shown to the right of the pore. This developmental stage is comparable to that shown in fig. 212 for *Curvularia inaequalis.* × 10,000.

Fig. 208

A sympodially proliferated conidiogenous cell of *D. sorokiniana* with a mature conidium (Cd) still attached. × 2,800.

Fig. 209

A frozen-etched conidium of *D. sorokiniana* showing the basal abscission scar (AS) and irregular surface of the wall. × 5,150.

Figs. 210–15

35-mm time-lapse photomicrographic sequence of a sympodially proliferated conidiogenous cell of *Curvularia inaequalis* (Shear) Boedijn. Three conidia have been produced in acropetal succession. Micrographs taken at 0, $1\frac{1}{4}$, 3, $3\frac{3}{4}$, 4 and $4\frac{3}{4}$ hours respectively. × 1,100.

Fig. 216

Section through part of a young conidium of *Drechslera sorokiniana.* The conidium is surrounded by a two layered wall (W1, 2) and contains several nuclei (N), vacuoles (V), lipid bodies (l), microbodies (mb) and a lomasome (L). Glutaralde-hyde–osmium tetroxide fixation. × 18,900.

Figs. 207–16 *207–9 From* COLE, G. T. (1973). *Can. J. Bot.,* **51**, 629. *210–15 From* COLE, G. T. (1971). *Taxonomy of Fungi Imperfecti* (ed. B. Kendrick), pp. 141–55. University of Toronto Press. *216 Micrograph by* DR G. T. COLE, University of Texas, Austin, Texas.

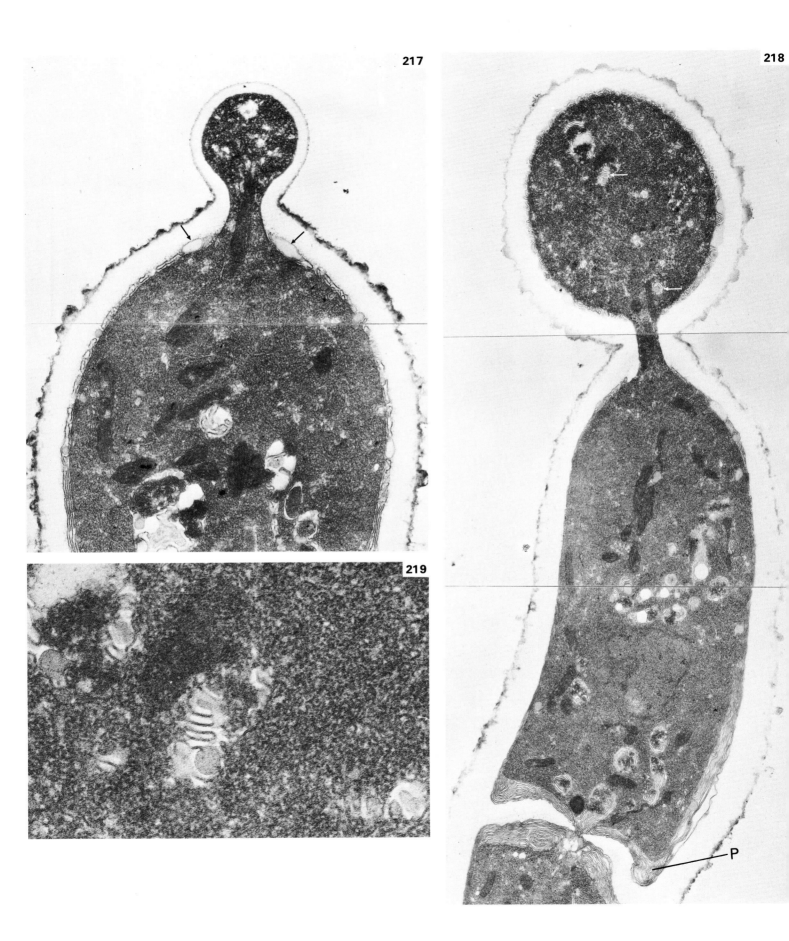

B. Reproductive Structures (cont.)
Asexual reproduction II (cont.)

Synchronous development of conidia in *Stemphylium botryosum* Wallr. can be induced by exposure of colonies grown at 27°C in darkness to a 12-hour period of near u.v. light (=320 − 420 nm) followed by return to darkness.

In the earliest stages observed the **conidial initial** can be seen as a minute bud whose wall is continuous with that of the conidiogenous cell (fig. 217). As the conidial initial enlarges areas of membrane convolution become visible in the cytoplasm of both the conidiophore and the conidium. Within these convolutions spherical, lightly staining bodies arise between the individual leaflets of a unit membrane (figs. 218, 219). Profiles of these bodies occur adjacent to the plasma membrane, suggesting cytoplasmic synthesis and secretion of some specific wall component. As cell maturation proceeds, first in the conidiogenous cell and then in the conidium, a zone of extensive convolution of the plasma membrane moves up in the conidiogenous cell **acropetally** (fig. 218). As the conidium enlarges and becomes septate, the cytoplasm within becomes very dense, and scattered non-staining floccose material appears (fig. 220). In such conidia differentiation of the wall into an outer **primary layer** surrounding the entire conidium and inner **secondary layers**, each surrounding a single cell within the primary layer (fig. 221) can be seen. During conidium maturation a radially aligned array of **fibrils** arises adjacent to the cytoplasmic neck joining conidiogenous cell and conidium (figs. 223–225) and a dense circular zone of pigment accumulates in the lateral wall of the conidiogenous cell (fig. 226). In the late stages of conidium maturation the apical cell of the conidiophore senesces and finally dies (fig. 226). A Woronin body remains closely appressed to the rim of the pore within the conidium (fig. 223) and presumably prevents migration of organelles as the conidiophore and conidium pursue their disparate developmental pathways.

Additional reading

CARROLL, FANNY E. and CARROLL, G. C. (1971). Fine structural studies on 'poroconidium' formation in *Stemphylium botryosum*. In: *Taxonomy of Fungi Imperfecti* (ed. B. Kendrick), pp. 75–91. University of Toronto Press.
CARROLL, FANNY E. (1972). A fine-structural study of conidium initiation in *Stemphylium botryosum* Wallroth. *J. Cell Sci.*, **11**, 33.

Fig. 217
Longitudinal section through a young conidium initial of *Stemphylium botryosum* Wallr. (17 hours after u.v. treatment). Masses of amorphous, lightly staining substance (arrows) occur adjacent to the wall at the base of the cytoplasmic junction between the conidial initial and the conidiogenous cell. × 24,000.

Fig. 218
Conidiophore bearing a young conidium (17 hours). Note convolution of the plasma membrane towards the base of the conidiogenous cell (P) and production of small spherical grey bodies in association with areas of localized membrane convolution in the conidium initial (arrows). × 16,000.

Fig. 219
Detail of fig. 218 showing formation of a grey body between the leaflets of a single membrane. × 81,000.

Figs. 217–19 *Glutaraldehyde–osmium tetroxide fixation. From* CARROLL, FANNY E. (1972). *J. Cell Sci.*, **11**, 33.

220

221

222

W 2

W 1

B. Reproductive Structures (cont.)
Asexual reproduction II (cont.)

Fig. 220

A young conidium of *S. botryosum* (21 hours). The conidium has now become multiseptate, and floccose material has appeared throughout the cytoplasm. ×11,000.

Fig. 221

A mature conidium (60 hours). The cytoplasm has become very dense, and the cells are filled with lipid and the flocculent substance noted in fig. 220. Note primary wall layer (W1) and secondary wall layers (W2). ×8,000.

Fig. 222

Detail of fig. 221 showing stratification of secondary wall. ×19,000.

Figs. 220–2 *Glutaraldehyde–osmium tetroxide fixation. Micrographs by* DR FANNY CARROLL, University of Oregon, Eugene, Oregon.

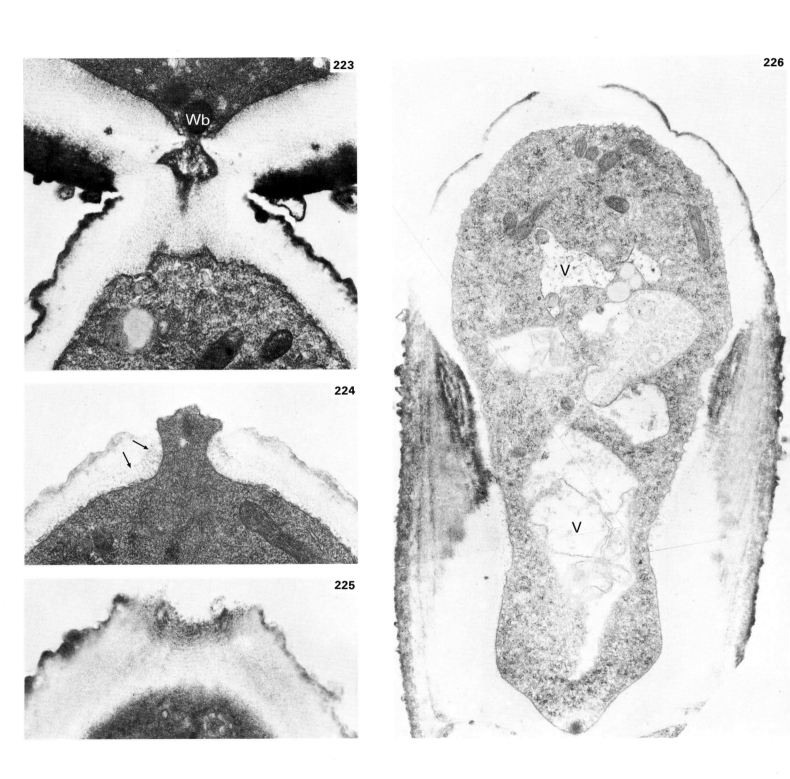

B. Reproductive Structures (cont.)
Asexual reproduction II (cont.)

Fig. 223

Longitudinal section through the junction between a maturing conidium and conidiophore (21 hours). Note the Woronin body (Wb). × 35,000.

Fig. 224

Tip of conidiogenous cell (21 hours). The conidium has been dislodged. Note fibrils in the wall surrounding the cytoplasmic isthmus leading to the conidium (arrows). × 35,000.

Fig. 225

Oblique section of conidiogenous cell tip (27 hours). Note concentric arrangement of fibrils in the wall below pore. × 35,000.

Fig. 226

Conidiogenous cell tip after senescence has begun (46 hours). Autophagic vacuoles (V) have begun to develop from swollen endoplasmic reticulum cisternae at many places throughout the cytoplasm. × 21,000.

Figs. 223–6 *Glutaraldehyde–osmium tetroxide. 223, 225–6 Micrographs by* DR FANNY CARROLL, University of Oregon, Eugene, Oregon. *224 From* CARROLL, FANNY E. and CARROLL, G. C. (1971) 'Fine structural studies on "Poroconidium" formation in Stemphylium botryosum', *Taxonomy of Fungi Imperfecti* (ed. B. Kendrick), pp. 75–91. University of Toronto Press.

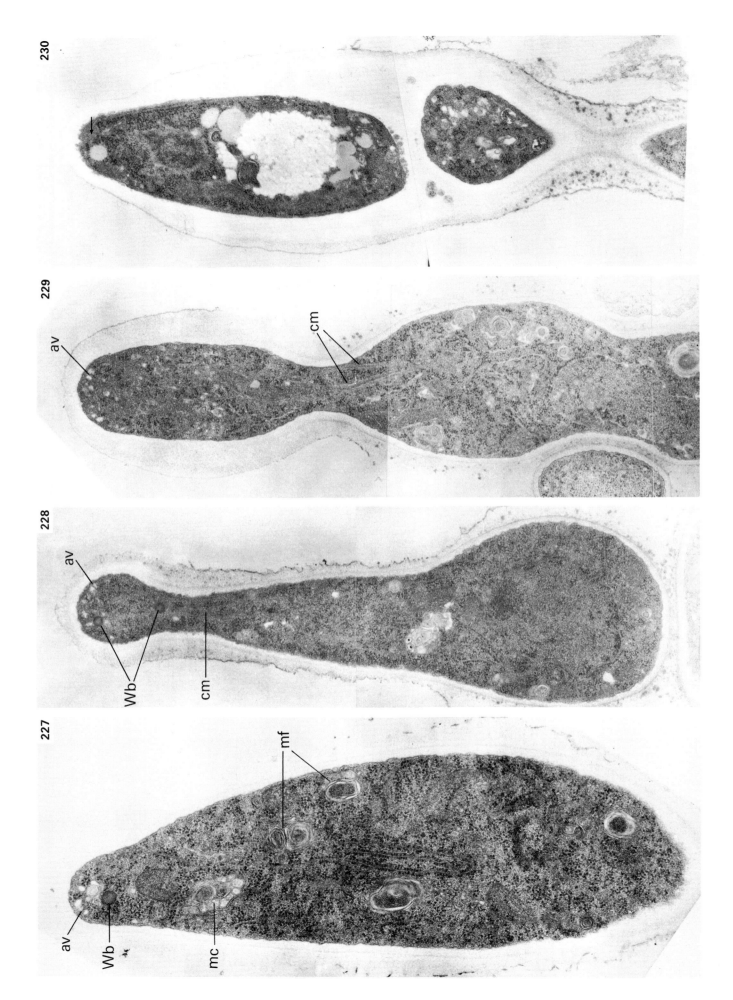

B. Reproductive Structures (cont.)
Asexual reproduction II (cont.)

In *Phialocephala dimorphospora* Kendrick synchronous development of conidia can be induced beneath a dialysis membrane overlay. After 48 hours' growth at 21°C the membrane is stripped from the culture and synchronous sporulation occurs. Some degree of asynchrony may however be found at all stages of development.

In the earliest stages of development a small accumulation of **apical vesicles** at the tip of the **phialide** marks the site of conidium initiation. Rough endoplasmic reticulum, convoluted membrane systems and myelin figures thought to be derived from mitochondria are apparent within the body of the phialide (fig. 227). As the conidium initial enlarges microtubules appear in the neck of the young phialide (fig. 228). Following nuclear division and migration of a single daughter nucleus into the spore, septum formation occurs. The cytoplasm of the maturing conidium becomes very dense, and inclusions presumed to be lipid and polysaccharide appear towards the interior of the spore (fig. 230). The inner wall of the **primary conidium** separates from the outer wall, which later ruptures as subsequently formed **secondary conidia** push the mature primary conidium forward (figs. 230, 231). The outer wall of the primary conidium thus becomes the **collarette** of the phialide. Completely mature conidia lose their electron density and reveal a rather simple internal organization. A single nucleus, several microbodies and mitochondria, numerous free ribosomes and presumed storage bodies are visible within each conidium. All traces of internal membrane systems and microtubules vanish. The plasma membrane becomes convoluted into a meandering network of grooves (figs. 231, 236).

During initiation of secondary conidia, apical vesicles are not apparent. However, large vesicles containing minute membrane fragments develop just below the neck of the phialide and appear to contribute to wall formation as they fuse with the plasma membrane, discharging their contents to the outside (figs. 232, 233). Such vesicles appear to arise through degradation of membrane from the previously noted convoluted membrane systems (figs. 227, 233, 235). As the secondary conidia are delimited by septa, they also become very electron dense and then rapidly lose this electron density as the grooves form in the plasma membrane (figs. 234, 236). Membrane complexes found in association with the plasma membrane at the base of each spore at this stage may be involved in the elaboration of additional plasma membrane stored in the grooves (fig. 235). Signs of senescence and organelle degeneration become apparent in phialides from cultures over one week old (fig. 237).

Fig. 227

Longitudinal section through a young phialide of *Phialocephala dimorphospora* Kendrick (18 hours after removing the membrane from the culture). Apical vesicles (av) and a Woronin body (Wb) are seen near the apex where initiation of the primary conidium occurs. Convoluted membranes (mc) and the myelin figures (mf) are also present. × 30,000.

Fig. 228

A young phialide (15 hours). The conidium initial has enlarged slightly. Woronin bodies (Wb) and apical vesicles (av) are found near the site of apical growth. Microtubules (cm) have appeared in the neck of the phialide. × 22,500.

Fig. 229

A young phialide (19 hours). The primary conidium has attained nearly full size. Apical vesicles (av) and cytoplasmic microtubules (cm) are visible. × 22,500.

Fig. 230

A young phialide (18 hours). The primary conidium has been delimited by a non-perforate septum and has been pushed forward by the first secondary conidium formed behind. A pocket of amorphous material at the very tip of the conidium (arrow) may be involved in the dissolution of the outer wall as the spore is released. × 22,500.

Figs. 227–30 *Glutaraldehyde–osmium tetroxide fixation. Micrographs by* DRS G. C. and FANNY E. CARROLL, Department of Biology, University of Oregon, Eugene, Oregon.

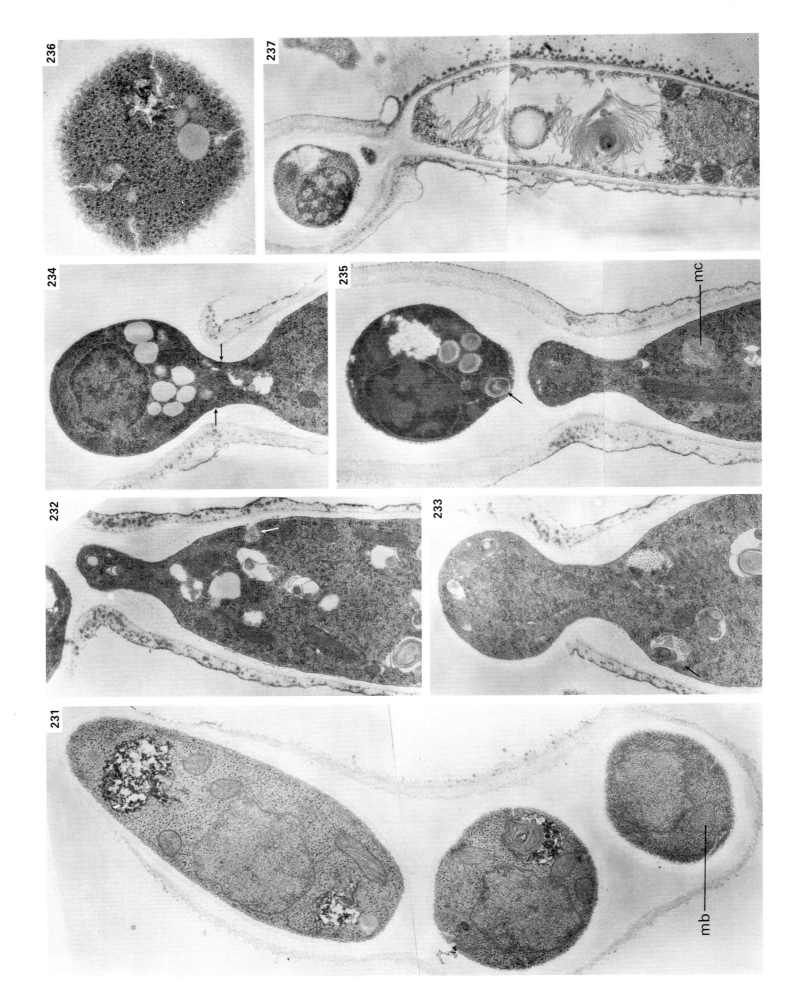

B. Reproductive Structures (cont.)
Asexual reproduction II (cont.)

Fig. 231

Longitudinal section through part of a phialide of *P. dimorphospora* showing three mature conidia still within the collarette (66 hours). The primary conidium (top) has just burst through the wall of the collarette. A microbody (mb) can be seen in the lower conidium. Glutaraldehyde/formaldehyde–osmium tetroxide fixation. ×22,500.

Fig. 232

The neck of a mature phialide showing an early stage in the formation of a secondary conidium (66 hours). Arrow shows a pocket of membrane fragments being released to the outside. ×22,500.

Fig. 233

The neck of a mature phialide showing a slightly later stage in secondary conidium formation (66 hours). The daughter nucleus has just migrated into the conidial initial. Arrow shows an intermediate stage in the degradation of a pocket of convoluted membrane. ×22,500.

Fig. 234

A later stage in the formation of a secondary conidium (66 hours). Lipid globules have appeared in the spore initial, and the formation of a basal septum has just begun (arrows). ×22,500.

Fig. 235

A late stage in the formation of a secondary conidium (66 hours). Myelin figure associated with the plasma membrane (arrow) is thought to be involved in the elaboration of grooves. Glutaraldehyde/formaldehyde–osmium tetroxide fixation. ×22,500.

Fig. 236

Tangential section of a mature secondary conidium showing grooves in the plasma membrane (66 hours). ×45,000.

Fig. 237

A senescent phialide (approx. 1 week). ×16,000.

Figs. 231–7 (*232–4, 236, 237 Glutaraldehyde–osmium tetroxide fixation.*) *231–7 Micrographs by* DRS G. C. and FANNY E. CARROLL, Department of Biology, University of Oregon, Eugene, Oregon.

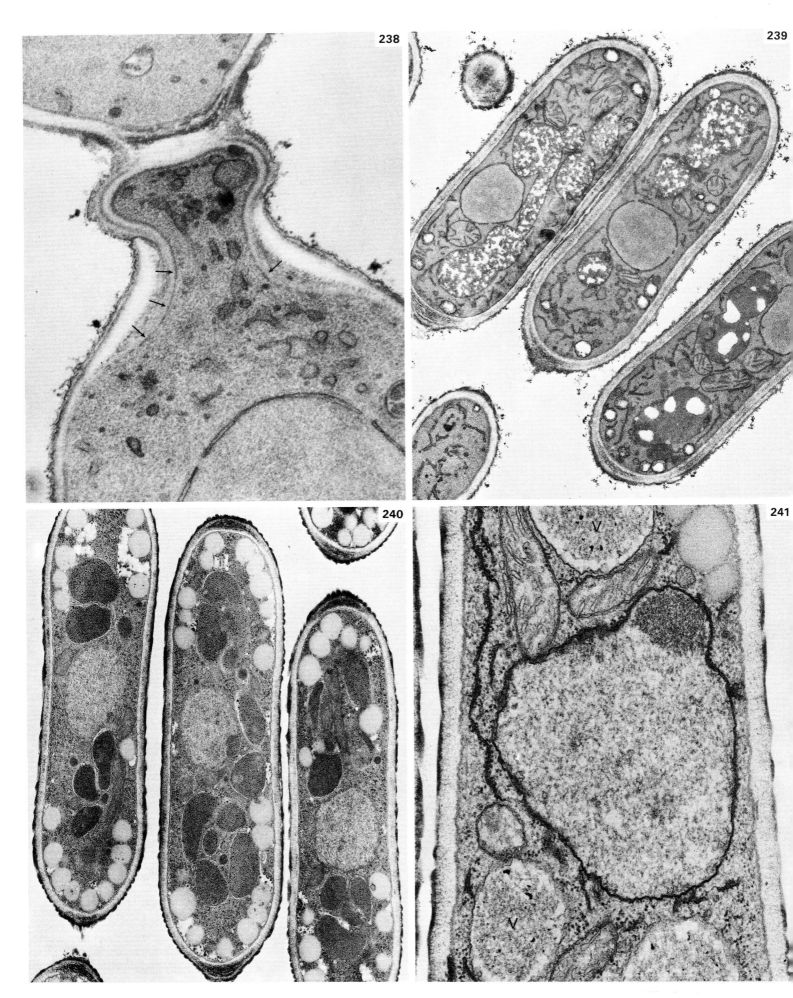

B. Reproductive Structures (cont.)
Asexual reproduction II (cont.)

The phialides of *Metarrhizium anisopliae* (Metsch.) Soro-kin are generally cylindrical to very slightly obclavate, hyaline and have very minute apices of less than 1·0 μm. A **basipetal** succession of conidia is produced from the apex of a **conidiogenous cell** the length of which remains constant. The phialide apex is distinct (fig. 238), and immediately behind it is a region of increased wall thickness. This swollen neck region has been shown in all published micrographs of phialides and is probably formed as a result of repeated conidium formation at a fixed locus (see additional reading on *Stachybotrys atra*, p. 117). After a conidium initial blows out, a septum is formed in the neck of the phialide. With subsequent conidium formation this septum moves distally out of the phialide, and eventually splits across the middle region so that half of the septum forms the base of one conidium and the other half the apex of the subsequent conidium (fig. 238). Layers of electron-opaque material are laid down adjacent to the septum prior to separation and will result in the formation of characteristic wall thickenings at the base and apex of conidia (fig. 240).

Conidia of *M. anisopliae* contain numerous vacuoles and lipid bodies. During conidium maturation the vacuoles undergo morphological changes as electron-opaque material accumulates within them (figs. 239, 240, 241).

Additional reading

HAMMILL, T. M. (1972). Electron microscopy of phialoconidiogensis in *Metarrhizium anisopliae. Am. J. Bot.*, **59**, 317.

Fig. 238

Longitudinal section through the tip of a phialide of *Metarrhizium anisopliae* (Metsch.) Sorokin, showing a young conidium initial emerging beneath a previously formed conidium. Note the layer of new wall material originating at the conidiogenous locus and adding to the thickening of the wall at the phialide apex (arrows). Electron-opaque lamellations can be seen in the wall at the base of the upper-most conidium. Formaldehyde—potassium permanganate fixation. × 45,500.

Figs. 239, 240

Maturing conidia of *M. anisopliae* showing the typical compli-ment of organelles and the variation in vacuolar contents. Note lamellations in the wall at the base and apex of the conidia. Formaldehyde—potassium permanganate fixation. × 13,500 and glutaraldehyde—osmium tetroxide fixation. × 13,800 respectively.

Fig. 241

Part of the central region of a young conidium at a stage prior to the accumulation of densely staining material within the vacuoles (V). Glutaraldehyde—osmium tetroxide fixation. × 58,000.

Figs. 238–41 *Micrographs by* DR T. M. HAMMILL, State University of New York, Oswego, New York.

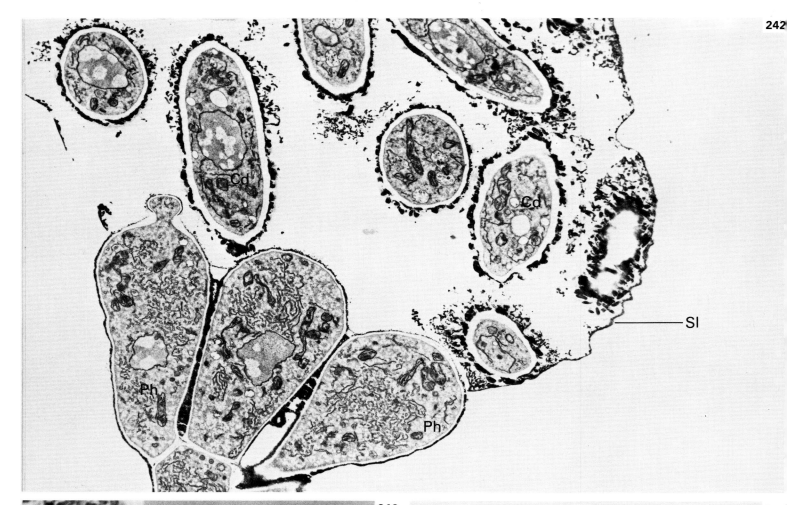

242

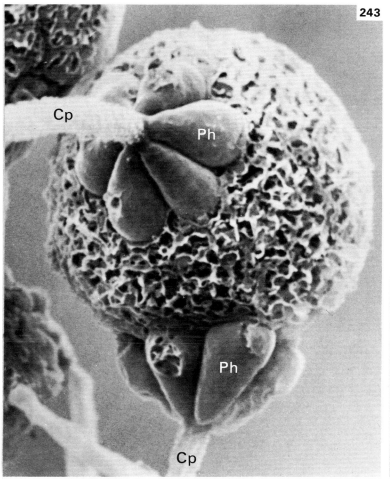

243

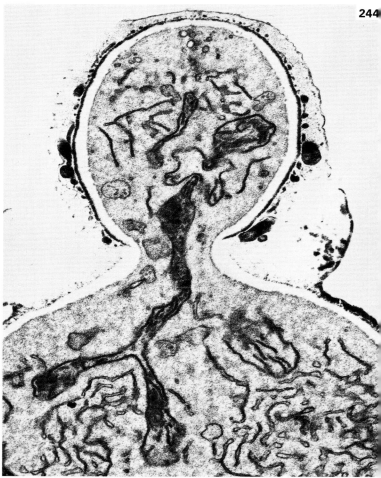

244

B. Reproductive Structures (cont.)
Asexual reproduction II (cont.)

Stachybotrys atra Corda bears its conidia in **slime drops** at the tips of clusters of phialides (figs. 242, 243). The active phialides contain one nucleus and an extensive endoplasmic reticulum system (figs. 242, 244, 245). The first conidium is produced by the expansion of the inner wall layer of the phialide (fig. 244), the slime is attached to the outside of the phialide tip. When the conidium reaches its final size it is cut off from the phialide by a septum formed by a centripetally growing invagination at the phialide neck (fig. 245). This septum becomes continuous across the neck and then splits, one half forming the base of the conidium and the other closing the phialide neck. The next conidium is formed by expansion of a new wall layer laid down inside the phialide neck. As the phialides age they therefore develop thickenings of the wall within the neck (figs. 242, 245) and these thickenings have a layered structure, each layer being the place of origin of one conidium. The phialide neck eventually becomes plugged by these thickenings or by electron dense material (fig. 247). The old phialides often have complex membrane configurations and two nuclei. The latter suggests that nuclear division may continue after conidium formation has stopped. The conidia (figs. 242, 246) have **truncated bases**, where they were attached to the phialides, and a two layered wall. The outer layer is electron dense and rough. Conidia contain mitochondria, vacuoles and endoplasmic reticulum (figs. 242, 246) and may also have prominent lipid droplets (fig. 246).

Additional reading

CAMPBELL, R. (1972). Ultrastructure of conidium ontogeny in the deuteromycete fungus *Stachybotrys atra* Corda. *New Phytol.*, **71**, 1143.

Fig. 242

Longitudinal section through a group of phialides (Ph) and conidia (Cd) of *Stachybotrys atra* Corda. Note slime sheath covering the conidia. Potassium permanganate fixation. × 9,200.

Fig. 243

Two conidiophores (Cp) with phialides (Ph) in a common slime drop. The rough surface of the drop is probably a drying artefact.

Freeze-dried, Au/Pd coated. Stereoscan electron micrograph. × 4,000.

Fig. 244

Section through the first conidium emerging from the phialide tip. Potassium permanganate fixation. × 31,500.

Figs. 242–4 *242, 244 From* CAMPBELL, R. (1972). *New Phytol.*, **71**, 1143. *243 Micrograph by* DR R. CAMPBELL, Department of Botany, Bristol University.

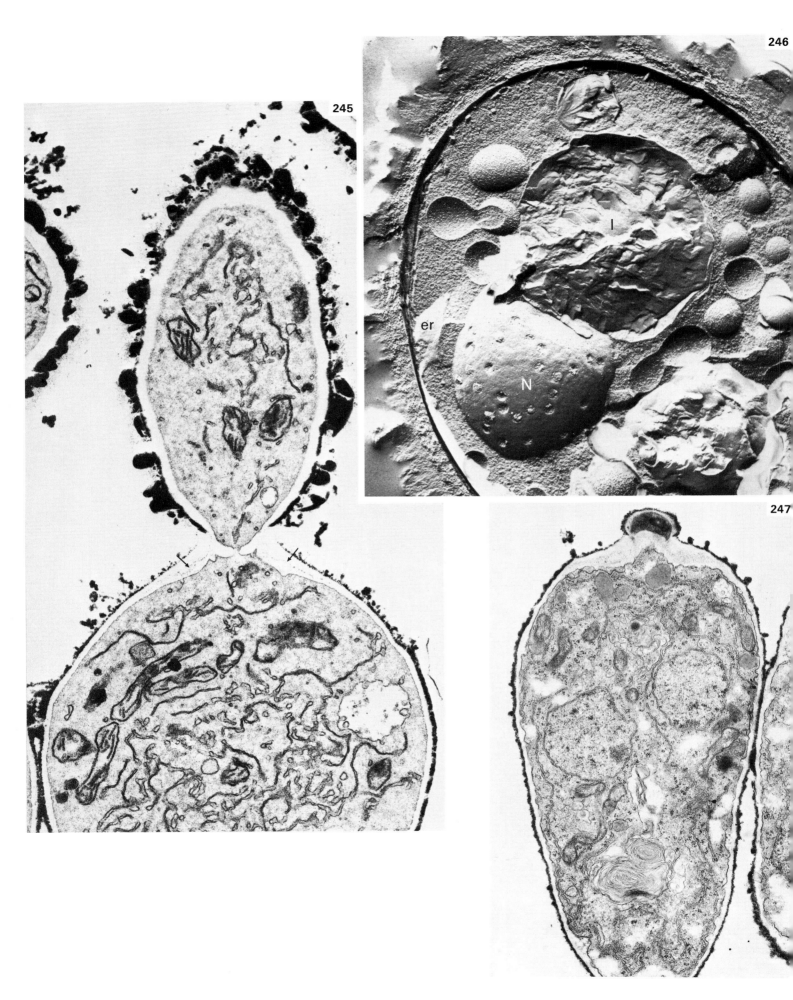

245

246

247

N

er

l

B. Reproductive Structures (cont.)
Asexual reproduction II (cont.)

Fig. 245

Longitudinal section through part of a phialide of *S. atra* with a fully developed conidium (the second produced from this phialide). The septum between conidium and phialide is invaginating and will eventually become complete. Note new wall layer (arrows) in the phialide neck which is the origin of the conidium wall. Potassium permanganate fixation. × 36,000.

Fig. 246

Freeze-etched preparation of a fully developed conidium. A nucleus (N) and large lipid body (I) can be seen together with the membrane surface of a sheet of endoplasmic reticulum (er). × 28,000.

Fig. 247

Longitudinal section through an old phialide that is no longer producing conidia. Glutaraldehyde/formaldehyde–osmium tetroxide fixation. × 14,700.

Figs. 245–7 *From* CAMPBELL, R. (1972). *New Phytol.*, **71**, 1143.

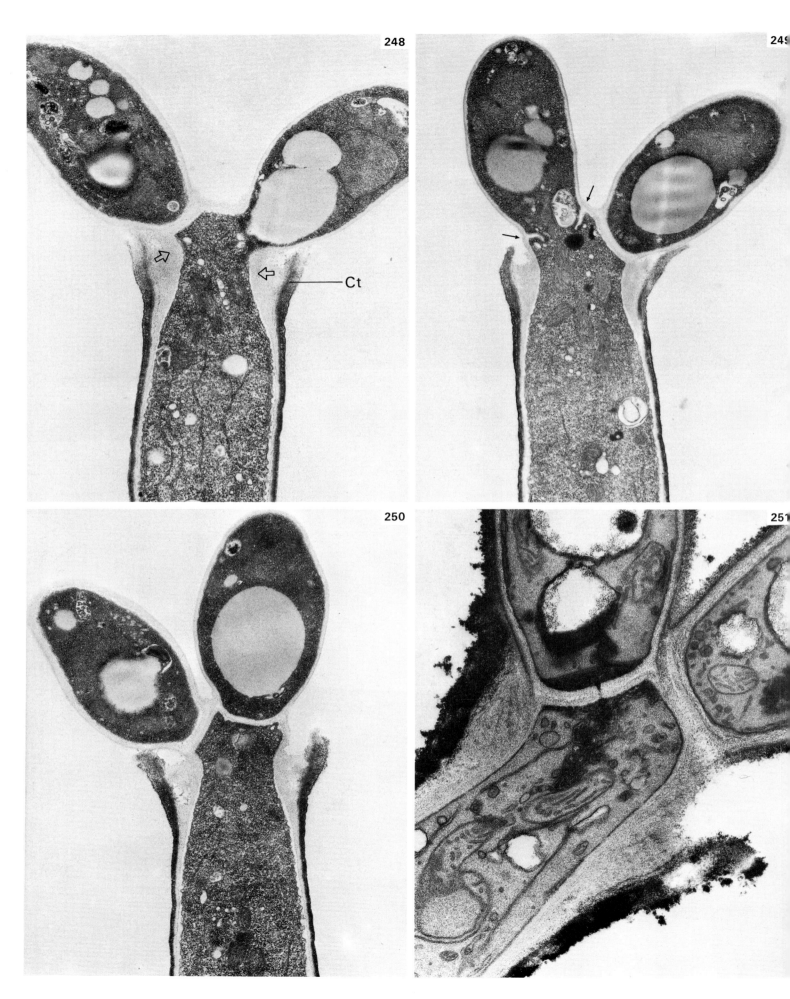

248

250

Ct

249

251

B. Reproductive Structures (cont.)
Asexual reproduction II (cont.)

The deuteromycete *Chloridium chlamydosporis* (van Beyma) Hughes produces a **sympodial** succession of conidia from successively new conidiogenous loci at the tip of what, from light microscopy, was thought to be a typical phialide (figs. 248–251). At the apex of the **conidiogenous cell** the outer wall is ruptured during **primary conidium** development to form a **collarette.** Once a conidium has developed, a **new conidium** initial forms by the expansion and synthesis of new wall at a different **conidiogenous locus** located beneath and to one side of the previously formed conidium. Repeated conidium production therefore leads to a thickening of the wall within the collarette (figs. 248–251). The conidiogenous cell in *C. chlamydosporis* is morphologically very similar to the typical phialide, but since conidia are produced sequentially from several different loci, then by definition it cannot be a phialide since the latter has a **fixed conidiogenous locus** (see additional reading, p. 103).

Additional reading

HAMMILL, T. M. (1972). Electron microscopy of conidiogenesis in *Chloridium chlamydosporis. Mycologia,* **64**, 1054.

Figs. 248–51

Longitudinal sections through parts of conidiogenous cells of *Chloridium chlamydosporis* (van Beyma) Hughes showing stages in the formation of conidia. Note the delimitation of conidia by centripetal cross wall formation (arrows) and the wall thickenings (open arrow) within the collarette (Ct). ×17,600; ×19,400; ×18,000; ×32,000; respectively.

Figs. 248–51 *(248–50 Glutaraldehyde–osmium tetroxide fixation. 251 Formaldehyde–potassium permanganate fixation.) 248–51 From* HAMMILL, T. M. (1972). *Mycologia,* **64**, 1054.

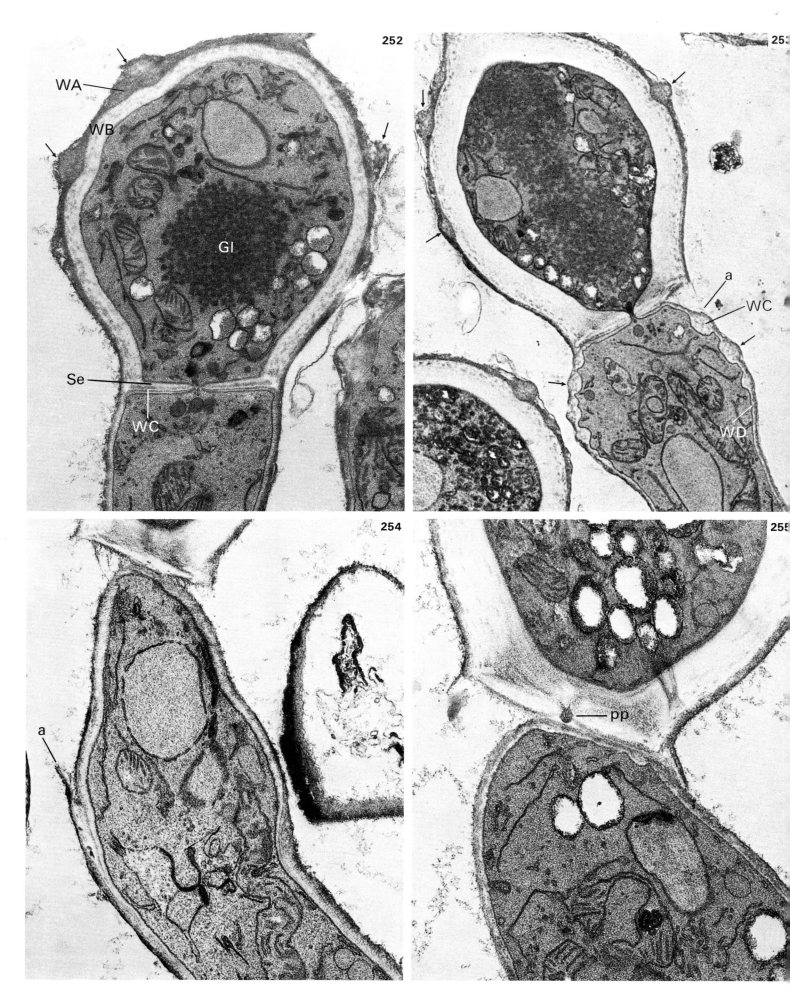

B. Reproductive Structures (cont.)
Asexual reproduction II (cont.)

In *Scopulariopsis brevicaulis* (Sacc.) Bain. and *S. koningii* (Oud.) Viull., conidium initials are delimited by centripetally growing septa (fig. 257) which eventually become bilayered (fig. 252). The upper layer of the septum increases in thickness with the inner wall layer (secondary wall) of the conidium (figs. 253, 260a). Localized thickenings of the outer, electron-opaque layer (primary wall layer) of the maturing conidium wall give rise to the characteristic **verrucose** nature of conidia in *S. brevicaulis* (figs. 252, 253, 256, 260a–f). When conidia are released an **annular scar** is usually left on the outside of the neck of the conidiogenous cell which is itself termed an **annellide** (figs. 253, 254, 258). As with phialides, the upper layer of the bilayered septum of the typical annellide forms the thick walled base of the released conidium and in *S. brevicaulis* and *S. koningii* remains continuous with the conidium secondary wall (figs. 254, 255, 260a, b). The lower layer of the septum forms part of the primary wall of the next conidium (fig. 260c–f). In both species a pore plug blocks the septal pore before conidia are completely released and this pore plug remains embedded in the base of the conidium (figs. 255, 260b–f). Frozen-etched annellides of *S. brevicaulis* show a characteristic rodlet pattern on the outer surface (figs. 258, 259).

Additional reading

COLE, G. T. and ALDRICH, H. C. (1971). Ultrastructure of conidiogenesis in *Scopulariopsis brevicaulis. Can. J. Bot.,* **49**, 745.
HAMMILL, T. M. (1971). Fine structure of annellophores I. *Scopulariopsis brevicaulis* and *S. koningii. Am. J. Bot.,* **58**, 88.

Figs. 252, 253

Sections through parts of annellides of *Scopulariopsis brevicaulis* (Sacc.) Bain. showing development of the septum (Se) at the base of the conidium, the localized wall thickenings (arrows), the annular scar (a) and the large deposits of glycogen (Gl). The wall layers (WA, WB, WC, WD) correspond to those shown in the diagram in fig. 260. ×22,000; ×11,600.

Figs. 254, 255

Parts of annellides of *S. koningii* (Oud.) Vuill. showing stages in the release of conidia. Note the annular scar (a) and septal pore plug (pp.). ×21,000; ×32,000.

Figs. 252–5 *Formaldehyde–potassium permanganate fixation. From* HAMMILL, T. M. (1971). *Am. J. Bot.,* **58**, 88.

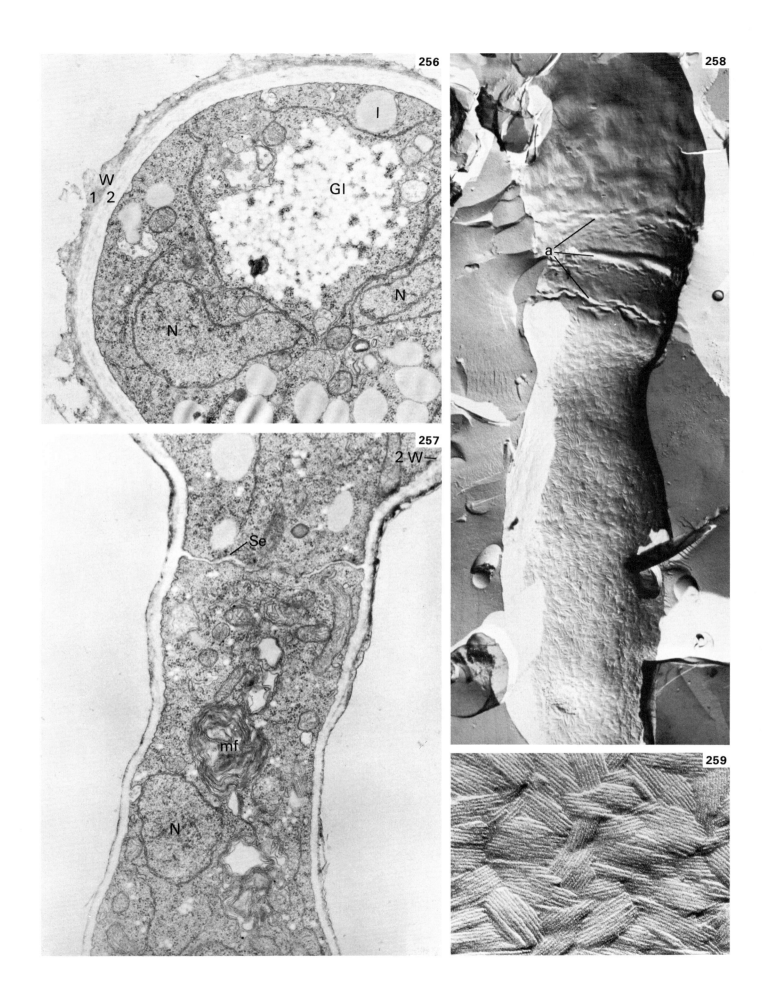

B. Reproductive Structures (cont.)
Asexual reproduction II (cont.)

Figs. 256, 257

Parts of a young conidium and conidiogenous cell of *S. brevi-caulis* showing the primary and secondary wall layers (W1, 2), a developing septum (Se), several nuclei (N), a myelin figure (mf), glycogen (Gl) and lipid bodies (l). Glutaraldehyde–osmium tetroxide fixation. ×19,800.

Figs. 258, 259

A frozen-etched conidiogenous cell of *S. brevicaulis* showing three annular scars (a) of previously formed conidia. The conidiogenous cell is terminated by a conidium initial. Note the pattern of rodlets on the wall surface. ×12,200; ×108,000.

Figs. 256–9 *From* COLE, G. T. and ALDRICH, H. C. (1971). *Can. J. Bot.,* **49**, 745–55. Reproduced by permission of the National Research Council of Canada.

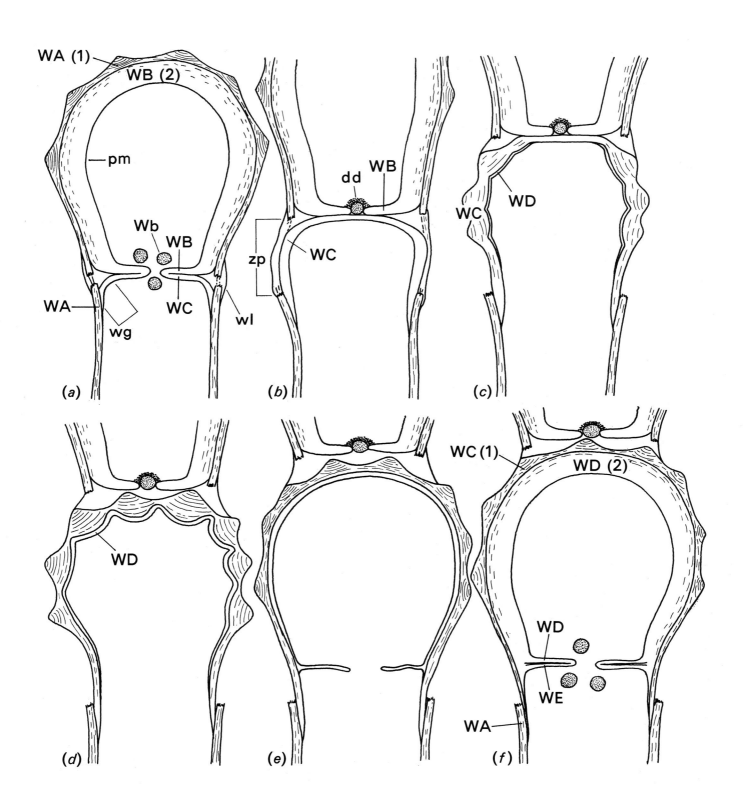

B. Reproductive Structures (cont.)
Asexual reproduction II (cont.)

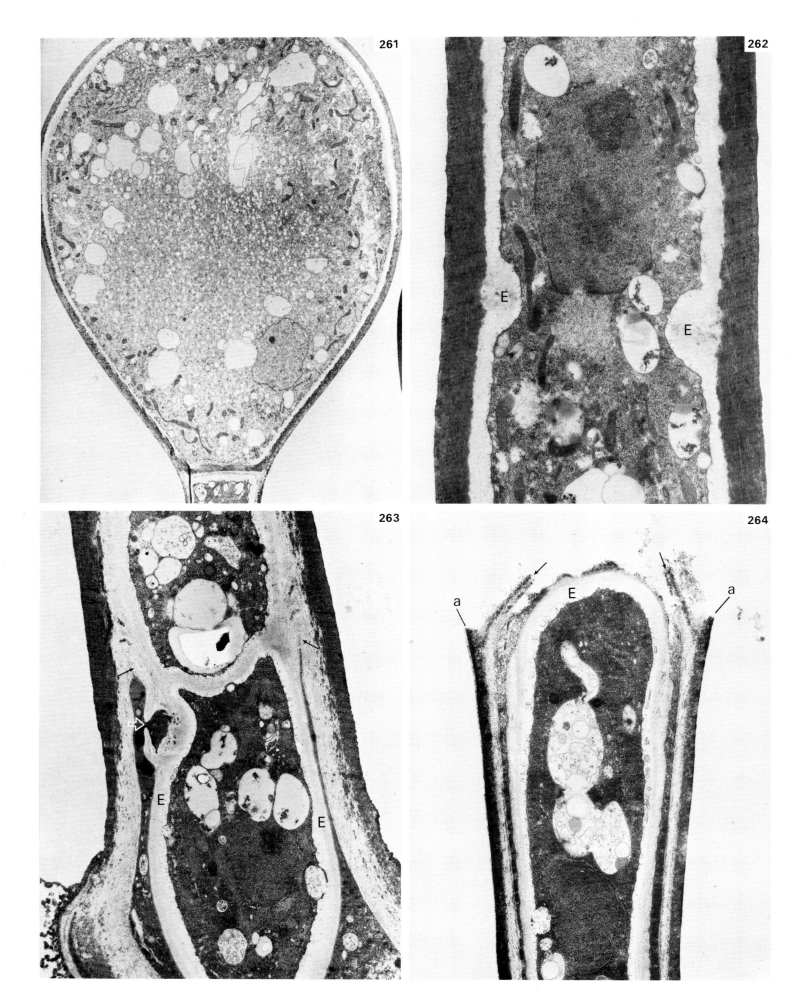

B. Reproductive Structures (cont.)
Asexual reproduction II (cont.)

Conidiogenous cells of *Monotosporella sphaerocephala* (Berk. and Br.) Hughes superficially resemble typical annellides, but they possess intraconidiophoral hyphae (**endohyphae**), which because of the dark pigmentation of conidiophore walls cannot be resolved with the light microscope. Endohyphae arise from the electron-transparent inner wall layer of the conidiophores (fig. 262) and grow within the conidiophore (figs 263, 264). Occasionally what appear to be **'false septa'** may be seen in unusual positions on endohyphae (fig. 263).

Young conidium initials in *M. sphaerocephala* are delimited by centripetally developing septa, and delimited conidia enlarge considerably. Conidium-delimiting septa develop an electron-opaque layer just below which is the plane of separation. The distal portion of a split septum forms the base of the released conidium; the proximal layer of a split septum forms the apex of the conidiogenous cell. Subsequent events however differ from those which occur in a 'typical' annellide. The proximal layer of the split septum does *not* form a part of the wall of either the proliferation or the next conidium. Instead, it is a passive barrier through which endohyphal proliferations grow. These rupture the apex of the conidiophore (fig. 264), and conidiogenesis follows the emergence of such endohyphal proliferations. Conspicuous annular scars form due to a rupturing of the electron-opaque outer wall layer of the conidiogenous cell. Conidia are forcibly separated from a conidiogenous cell by the growing endohyphae, or are displaced to one side. When the septal layers fail to split, the lower layer acts as a device for holding the displaced conidium to the side of the conidiophore.

Additional reading

HAMMILL, T. M. (1972). Fine structure of annellophores. III. *Monotosporella sphaerocephala. Can. J. Bot.*, **50**, 581.

Fig. 261
A young conidium of *Monotosporella sphaerocephala* (Berk. and Br.) Hughes with a thick, layered septum at the base. × 5,200.

Fig. 262
Longitudinal section through part of a conidiophore showing what is presumed to be the initiation of endohyphal development (E) by the invagination of the inner wall layer. × 16,600.

Fig. 263
Section through part of an endohypha (E) at the base of a conidiophore. Note fusion of endohypha with conidiophore inner wall layer (arrows) and the false septum (open arrow). × 11,500.

Fig. 264
Part of a conidiophore showing proliferation as a result of the penetration of an endohypha (E) through the remaining layer of septum (arrows), which had delimited the former conidium. Note annular scar (a). × 11,500.

Figs. 261–4 *Glutaraldehyde–osmium tetroxide fixation. From* HAMMILL, T. M. (1972). *Can. J. Bot.*, **50**, 581.

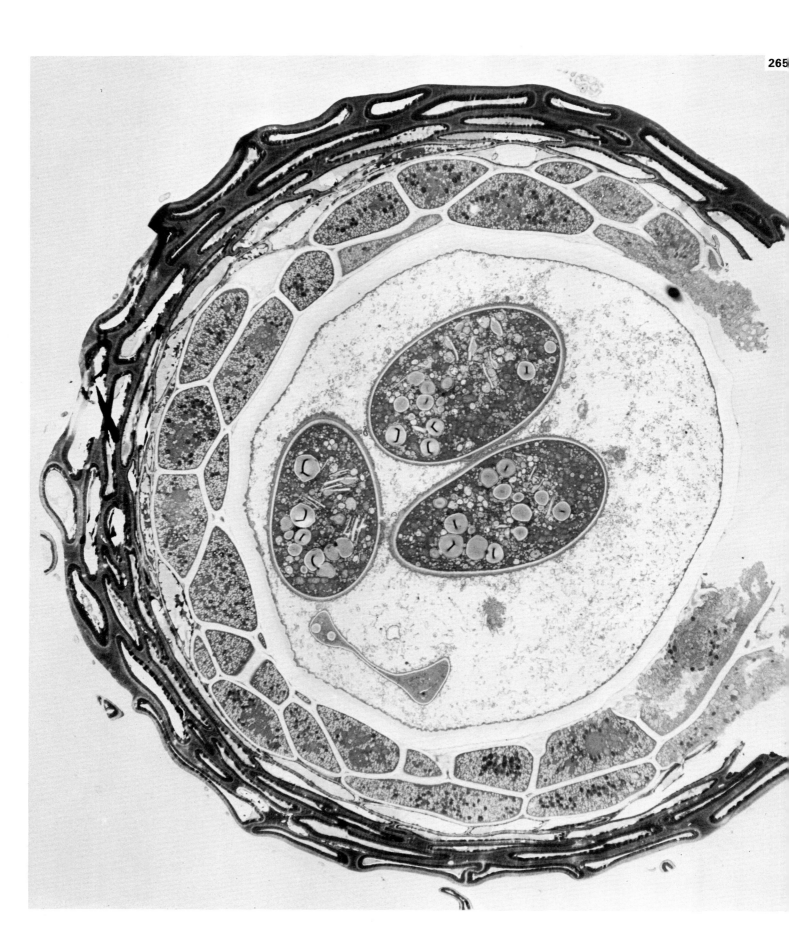

B. Reproductive Structures (cont.)
Sexual reproduction
Ascocarp structure

With the exception of the Hemiascomycetes, most Asco-mycotina produce some form of fruit body (the **ascocarp**) during sexual reproduction. Ascocarp morphology may vary considerably and modern concepts of Ascomycotina taxonomy are largely based on the development and structure of the fruit body and on the structure of the asci enclosed by them. Two of the recognized types of ascocarp will be dealt with here.

A comparatively simple form of ascocarp is the cleisto-thecium of *Sphaerotheca mors-uvae* (Schw.) Berk. and Curt. This is an enclosed globose structure, the wall of which is composed of several layers of highly melanized, dead cells (fig. 265). Within this hard outer coat is a layer of living cells which have a high lipid content. The large central cavity of the cleistothecium is entirely occupied by the **single**, thick-walled ascus, a characteristic of the genus *Sphaerotheca*.

The ascocarp of the Loculoascomycete *Stigmatea robertiani* Fr. differs structurally from the cleistothecium but more particularly in its mode of development. It is formed **subcuticularly** on the adaxial leaf surface of *Geranium robertianum*. Thick-walled, melanized hyphae lying between the cuticle and epidermis (fig. 266), swell and undergo cytokinesis. The maturing ascocarp, a **pseudothecium**, is a hemispherical structure composed largely of highly melanized cells (fig. 267). Within the pseudothecium a central cavity or **loculus** develops as a result of cell disintegration and asci and pseudoparaphyses form within the loculus.

Very few investigations have been made at the present time on ascocarp development and structure by means of electron-microscope techniques. However, recent work suggests that such a technique could provide useful information on this subject particularly for the taxonomist.

Additional reading

BLANCHARD, R. O. (1972). Ultrastructure of ascocarp development in *Sporormia australis. Am. J. Bot.*, **59**, 537.
BLANCHARD, R. O. (1972). Origin and development of ascogenous hyphae and pseudoparaphyses in *Sporormia australis. Can. J. Bot.*, **50**, 1725.

Fig. 265
Section through the cleistothecium of *Sphaerotheca mors-uvae* (Schw.) Berk. and Curt. The outer layers of dead cells enclose a layer of cells which contain many lipid bodies. The single ascus has a thick, electron-transparent wall and three of the normally eight ascospores are visible in this plane of section. The tear in the ascocarp wall on the right is due to preparation damage. Glutaraldehyde—osmium tetroxide fixation. × 2,800.

Fig. 265 *Micrograph by* M. MARTIN *and* DR J. GAY, *Imperial College, London.*

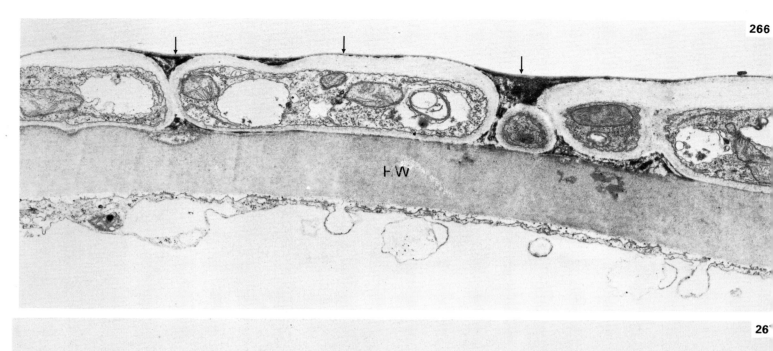

HW

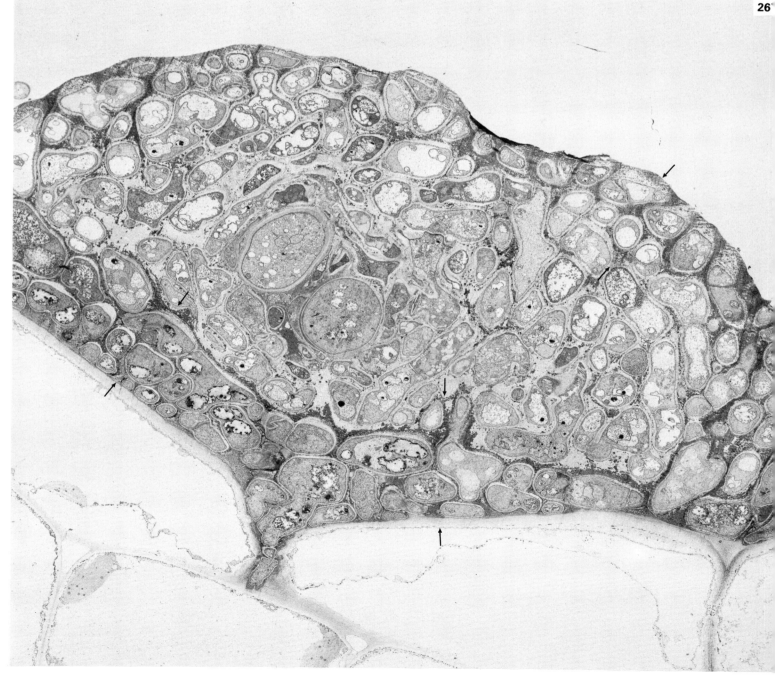

B. Reproductive Structures (cont.)
Sexual reproduction (cont.)
Ascocarp structure (cont.)

Fig. 266

Transverse section of a *Geranium robertianum* leaf showing the subcuticular hyphae of *Stigmatea robertiani* Fr. growing on the epidermal cell wall (HW) prior to ascocarp development. Note the thin cuticle (arrows), beneath which is a layer of electron-opaque material, probably melanin, which develops around the cells during ascocarp formation. × 14,000.

Fig. 267

Section through a developing ascocarp of *S. robertiani*. Note how the cells within the central region are relatively unmelanized, while the outer layer of 3–4 cells (between arrows) are embedded in a dense granular melanin-like substance. × 2,700.

Figs. 266, 267 *Glutaraldehyde–osmium tetroxide fixation. Micrographs by* A. BECKETT, Bristol University.

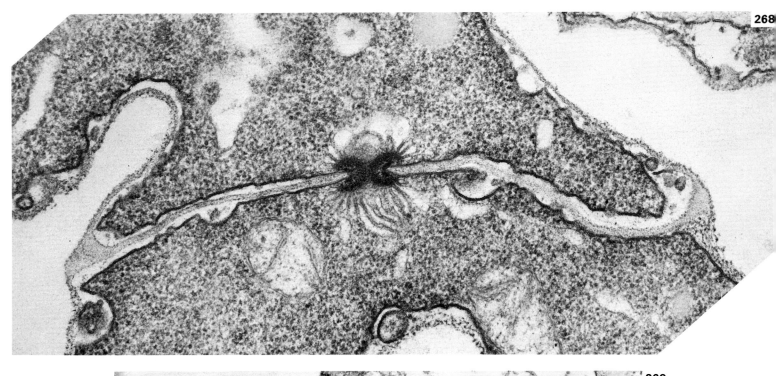

268

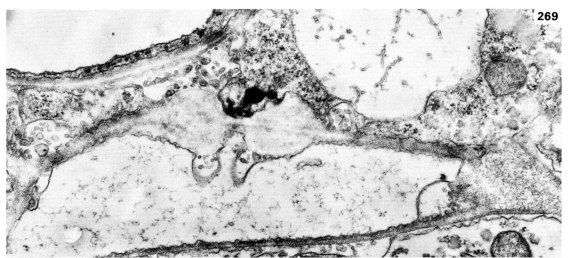

269

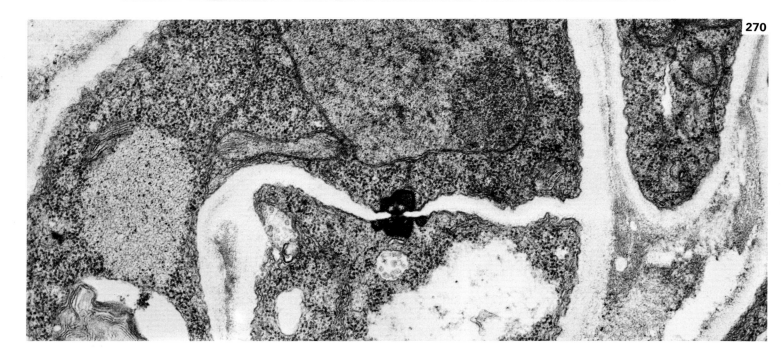

270

B. Reproductive Structures (cont.)

Sexual reproduction (cont.)

Septa and associated structures in ascogenous hyphae and asci

Recent work on the ultrastructure of developing asci has shown that in some cases elaborate structures are formed in association with the pore in the cross walls that delimit asci or ascogenous hyphae from subtending cells. These structures apparently vary in detailed morphology from genus to genus (figs. 268–270) and may also be species specific. Little is known of the function(s) of these structures but it would seem that they form a physical barrier to the passage of organelles, while soluble materials, needed for continued development of the ascus, could presumably pass through. In most, if not all fungi (see Section I, p. 51; Section II, p. 137; Section III, p. 204), spore-producing cells are cut off by septa and it may be that this delimitation from the vegetative mycelium is a fundamental requirement for the further development and functioning of what is essentially a specialized reproductive protoplast with a unique nuclear programme for producing spores. Three examples are illustrated.

In ascogenous hyphae and at the base of the ascus of *Sordaria macrospora* Auersw. the rim of the cross wall at the edge of the pore is lined by a swollen, dense ring from which tubular cisternae radiate into the cytoplasm of the two cells (figs. 268). A narrow pore perforates the centre of the ring.

In ascogenous hyphae of *Xylaria polymorpha* (Pers.) Grev. the edge of the cross wall surrounding the pore is inflated to approximately five times the thickness of the rest of the wall, to form an annular swelling (fig. 269) through which passes a narrow pore (seen only in median sections). The thin electron-transparent septal plate extends through the centre of the swelling and terminates at the edge of the pore (fig. 269). Woronin bodies frequently lie in the depression between the sides of the annular swelling, and are presumably capable of blocking the pore.

The pore apparatus in the septum at the base of the ascus of *Stigmatea robertiani* Fr. (fig. 270) consists of an electron-dense structure which, according to the fixative used, appears to be bound by a membrane and sometimes exhibits a substructure within it. Cisternae which are continuous with endoplasmic reticulum are often associated with this pore apparatus particularly on the ascus side of the pore.

Additional reading

BRACKER, C. E. (1967). Ultrastructure of fungi. *A. Rev. Phytopathol.*, **5**, 343.
CARROLL, G. C. (1967). The fine structure of the ascus septum in *Ascodesmis sphaerospora* and *Saccobolus kerverni*. *Mycologia*, **59**, 527.
FURTADO, J. S. (1971). The septal pore and other ultrastructural features of the pyrenomycete *Sordaria fimicola*. *Mycologia*, **63**, 104.

Fig. 268

Median longitudinal section through the septum and pore apparatus in an ascogenous hypha of *Sordaria macrospora* Auersw. Glutaraldehyde–osmium tetroxide fixation. × 60,000.

Fig. 269

Non-median section through the septum and pore apparatus in an ascogenous hypha of *Xylaria polymorpha* (Pers.) Grev. Glutaraldehyde–osmium tetroxide fixation. × 25,000.

Fig. 270

Median longitudinal section through the septum and pore apparatus at the base of an ascus of *Stigmatea robertiani* Fr. Glutaraldehyde–osmium tetroxide fixation. × 34,500.

Figs. 268–70 *268 Micrograph by* DR DENISE ZICKLER, Laboratoire de Génétique, Orsay. *269, 270 Micrographs by* A. BECKETT, Bristol University.

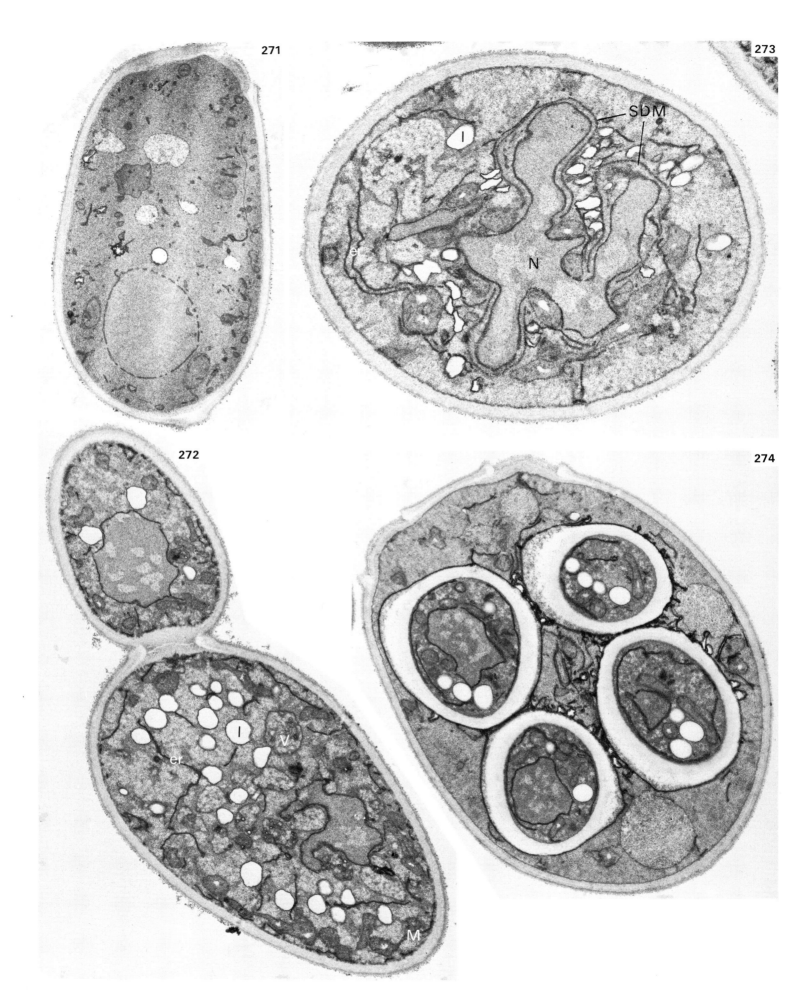

B. Reproductive Structures (cont.)

Sexual reproduction (cont.)

Ascus development

Sexual reproduction in all Ascomycotina involves the production of endogenous spores within specialized cells known as **asci**. In the yeast *Saccharomyces cerevisiae* Hansen any diploid cell is capable of becoming an ascus when grown under suitable conditions. During ascus development and maturation certain marked changes take place in the fine structure of the cell contents (figs. 271–273). In the ascus (as compared with the undifferentiated diploid cell, fig. 271) there are many more membranes, vesicles, vacuoles and lipid bodies (fig. 272). The cell itself enlarges and the nucleus divides (fig. 273), and finally ascospores are produced (fig. 274). During the formation of ascospore walls, certain inclusions of the ascus **epiplasm** (particularly lipids) are utilized.

In the Hemiascomycete *Taphrina deformans* (Berk.) Tul. asci are formed from diploid cells which lie beneath the cuticle of the host leaf *Prunus amygdalus* (fig. 275). Elongation of the asci causes the cuticle to rupture (figs. 276, 279), and the thick, two-layered ascus wall becomes stretched around the apex of the extending ascus (fig. 277). At the base of the ascus a cross wall forms by centripetal growth from the lateral ascus wall. This septum delimits the **basal** cell from the ascus proper. The cytoplasm of the basal cell vacuolates and aborts (fig. 277). Prior to ascus dehiscence, ascospores are located in the upper region of the ascus (fig. 278, 280), due to the formation of a large basal vacuole (fig. 278). Mature asci extend beyond the leaf surface and discharge ascospores through a characteristic slit in the wall at the ascus tip (fig. 281).

In the Pyrenomycetes asci typically develop from dikaryotic **ascogenous hyphae** which form **croziers**. Normally nuclear fusion occurs within the **penultimate cell** of the crozier to form an **ascus initial** (fig. 283). Elongation of the ascus initial involves cell wall and plasma membrane synthesis and this may be manifested at the ultrastructural level by the presence of **lomasomes** (figs. 283, 285, 286). Basal and apical vacuoles also develop in elongating asci (figs. 284, 286) and in some species, for example *Rosellinia aquila* (Fr.) de Notaris and *Xylaria longipes* Nitschke, ascus elongation is accompanied by a localized wall synthesis at the tip to form a complex **apical ring** (figs. 285, 286). Characteristic **apical vesicles** and a **Spitzenkörper** are involved in the development of this ascus apical apparatus (fig. 285).

Additional reading

BECKETT, A. and CRAWFORD, R. M. (1973). The development and fine structure of the ascus apex and its role during spore discharge in *Xylaria longpipes. New Phytol.,* **72**, 357.

CAMPBELL, R. (1973). Ultrastructure of asci, ascospores and spore release in *Lophodermella sulcigena* (Rostr.) v Hohn. *Protoplasma,* **78**, 69.

ILLINGWORTH, R. F., ROSE, A. H. and BECKETT, A. (1973). Changes in the lipid composition and fine structure of *Saccharomyces cerevisiae* during ascus formation. *J. Bact.,* **113**, 373.

WELLS, K. (1972). Light and electron microscopic studies of *Ascobolus stercorarius* II. Ascus and ascospore ontogeny. *University of California Publications in Botany,* **62**, 1. University of California Press.

Fig. 271

Section through an undifferentiated diploid cell of *Saccharomyces cerevisiae* Hansen. × 15,000.

Fig. 272

A developing ascus of *S. cerevisiae* showing numerous lipid bodies (l), vacuoles (V), mitochondria (M) and much endoplasmic reticulum (er). × 14,000.

Fig. 273

An ascus of *S. cerevisiae* in which the nucleus (N) is undergoing meiosis and the spore-delimiting membranes (SDM) are forming around what are presumed to be the poles of the meiotic spindle. (KMnO$_4$ does not preserve spindle microtubules. See figs. 310–312.) Note the development of endoplasmic reticulum (er) and lipids (l). Potassium permanganate fixation. × 16,000.

Fig. 274

A mature ascus containing four ascospores. Note reduction of lipid bodies and membranes in the ascus epiplasm. × 16,000.

Figs. 271–4 *Potassium permanganate fixation. 271–2, 274 Micrographs by* A. BECKETT, *Bristol University. 273 From* ILLINGWORTH, R. F., ROSE, A. H. *and* BECKETT, A. (1973). *J. Bact.,* **113**, 373

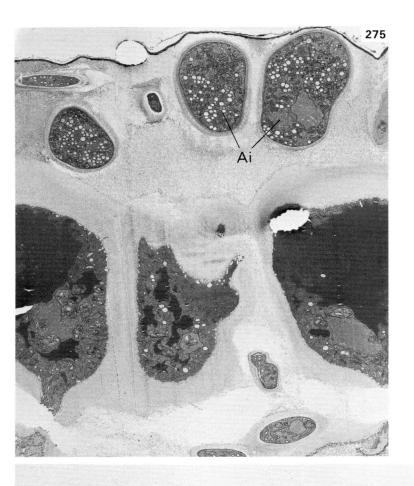

275

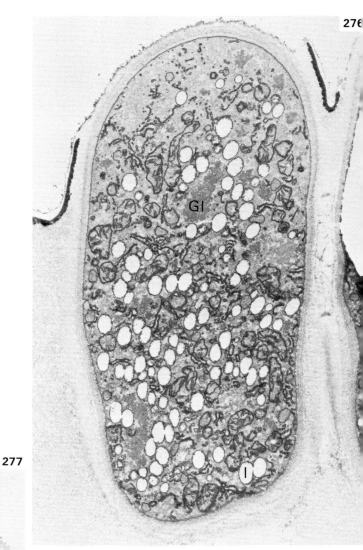

276

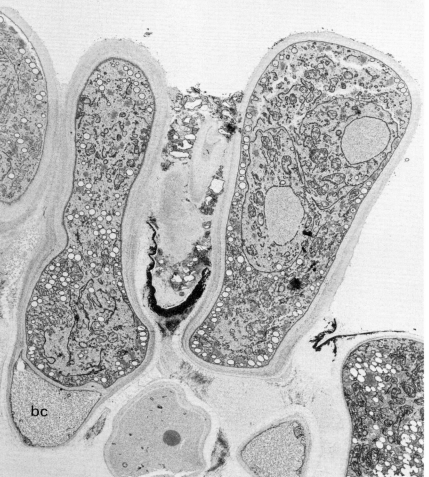

277

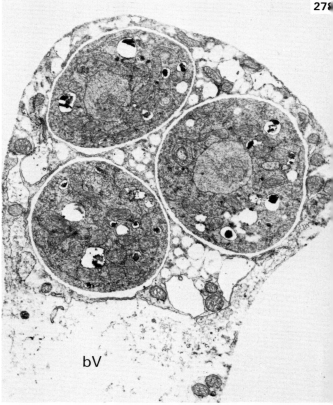

278

B. Reproductive Structures (cont.)
Sexual reproduction (cont.)
Ascus development (cont.)

Fig. 275

Transverse section through part of a leaf of *Prunus amygdalus* showing ascus initials (Ai) of *Taphrina deformans* (Berk.) Tul. beneath the cuticle. Note the mass of lipid bodies, endoplasmic reticulum and a large diploid nucleus in one initial. × 3,500.

Fig. 276

Longitudinal section through a developing ascus which is just emerging through the ruptured cuticle. Asci at this stage typically contain many lipid bodies (I) and large masses of glycogen (Gl). × 11,000 (compare with fig. 279).

Fig. 277

Longitudinal section through two elongated asci of *T. deformans* showing the formation of the basal cell (bc) in one of them, the two-layered ascus wall, and the typical 'spatulate' shape of the fully formed ascus at the stage of ascospore delimitation. × 3,900.

Fig. 278

Section through the upper part of an ascus in which there are three of the initially eight ascospores. Note how the basal vacuole (bV) forces the spores into the apical region of the ascus causing it to bulge. Glutaraldehyde/formaldehyde—osmium tetroxide fixation. × 11,250 (compare with fig. 280).

Figs. 275–8 (*275–7 Potassium permanganate fixation*.) *275–8 Micrographs by* MARY SYROP, Department of Botany, Bristol University.

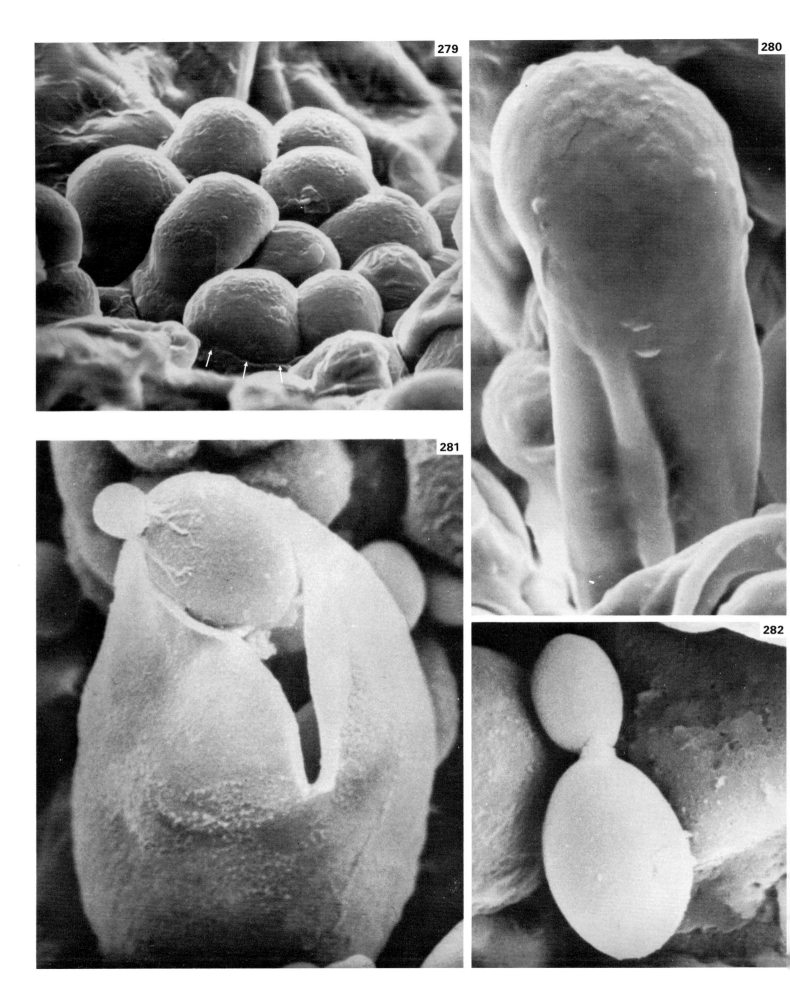

B. Reproductive Structures (cont.)
Sexual reproduction (cont.)
Ascus development (cont.)

Fig. 279

Stereoscan electron micrograph of developing asci of *T. deformans* emerging through the ruptured cuticle of the host leaf. Note edge of broken cuticle (arrows). × 4,700.

Fig. 280

Stereoscan electron micrograph of an elongated ascus of *T. deformans* showing the swollen upper portion due to the presence of spores. The lower vacuolated region has partially collapsed during preparation. × 9,100.

Fig. 281

Stereoscan electron micrograph of a dehiscing ascus. A budding ascospore rests in the gaping slit of the ascus wall. × 8,400.

Fig. 282

A discharged budding ascospore. × 11,000.

Figs. 279–82 *Micrographs by* MARY SYROP. Department of Botany, Bristol University.

Fig. 283

Longitudinal section through an ascus initial of *Xylaria longipes* Nitschke showing lomasomes (L) in the growing region, rough endoplasmic reticulum (rer) and numerous cytoplasmic ribosomes (in circle). Glutaraldehyde/acrolein—osmium tetroxide fixation. × 13,400.

Fig. 284

Longitudinal section through a young ascus of *Lophodermella sulcigena* (Rostr.) v Hohn. showing basal and apical vacuoles (bV) (aV) below and above the diploid nucleus. Glutaraldehyde/formaldehyde—osmium tetroxide fixation. × 4,400.

Fig. 285

Longitudinal section through the apical region of an elongating ascus of *Rosellinia aquila* (Fr.) de Not. in which the apical ring is partially formed (compare with fig. 293). Note the apical body or Spitzenkörper (ab) and numerous vesicles beneath the apical ring, and the lomasomes (L) in the wall. Glutaraldehyde—osmium tetroxide fixation. × 25,000.

Fig. 286

Tangential longitudinal section through part of an elongating ascus of *Xylaria longipes* Nitschke showing an early stage in apical vacuole formation (aV), the meiotic prophase nucleus (N) and numerous lomasomes (L) in the cell wall. Note how the microtubules run through the cytoplasm parallel to the ascus long axis near to the cell wall (e.g. where the plane of section passes close to cell wall the ascus is narrow and numerous microtubules are visible). Several mitochondria are closely associated with microtubules (arrows). Glutaraldehyde/acrolein—osmium tetroxide fixation. × 10,000.

Figs. 283–6 *283 From* BECKETT, A. and CRAWFORD, R. M. (1973). *New Phytol,* **72**, 357. *284 From* CAMPBELL, R. (1973). *Protoplasma,* **78**, 69. *285–6 Micrographs by* A. BECKETT, Bristol University.

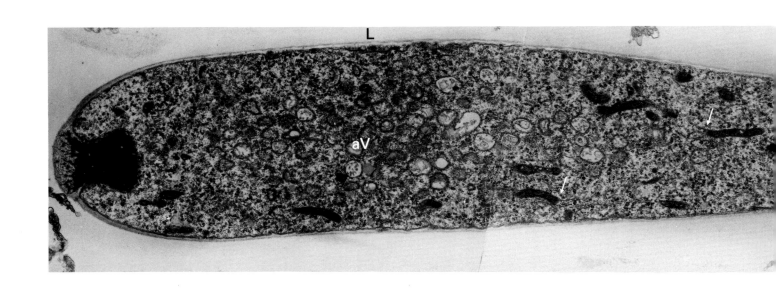

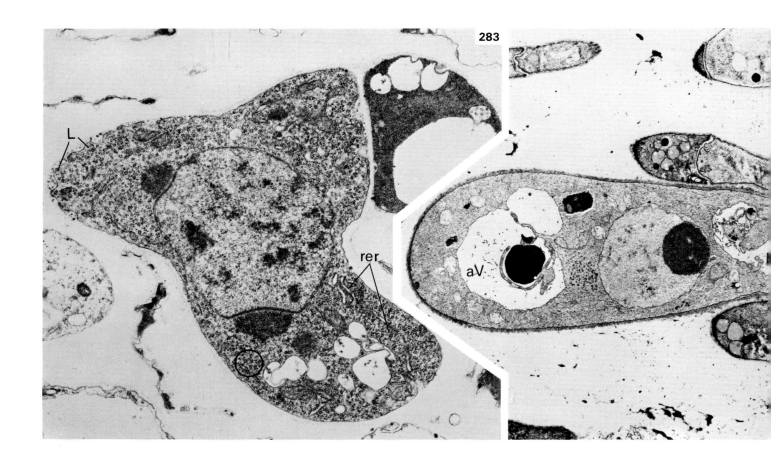

283

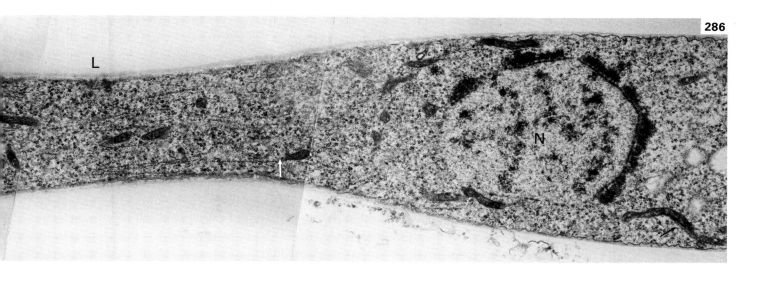

286

L

N

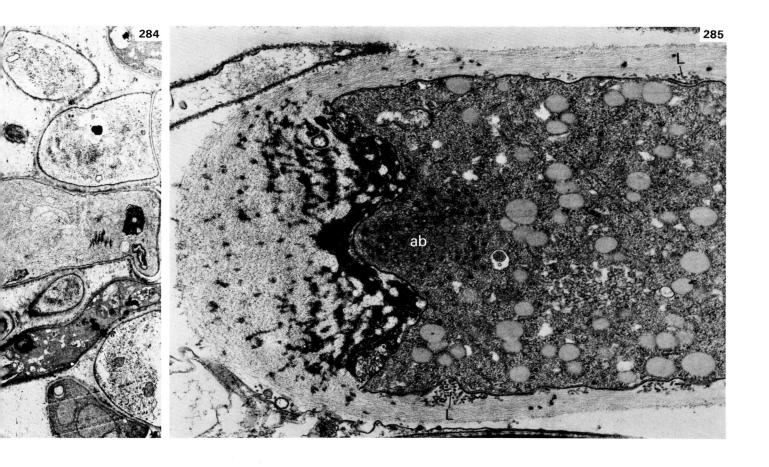

284

285

L

ab

L

L

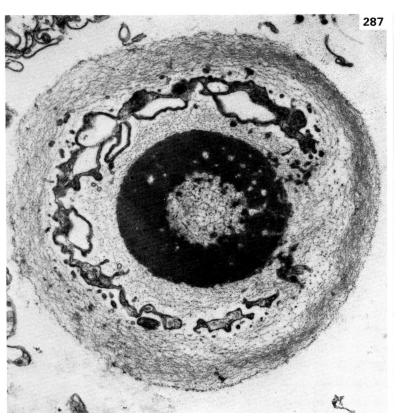

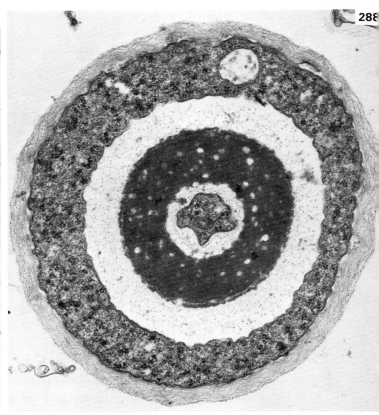

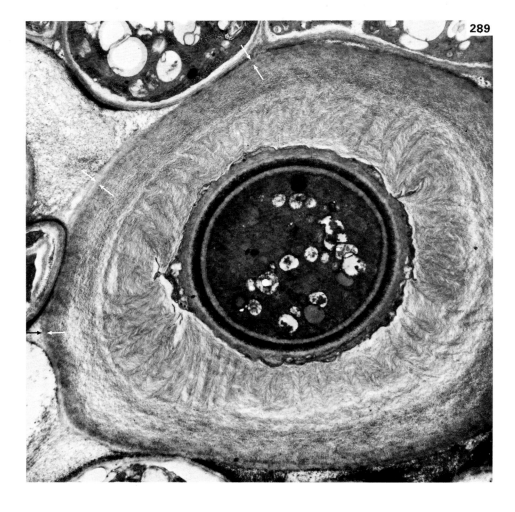

B. Reproductive Structures (cont.)
Sexual reproduction (cont.)
Ascus wall structure

Asci may be distinguished as **unitunicate** or **bitunicate** on the basis of their wall structure and mode of dehiscence. The division into these two categories is now considered to be of fundamental importance by most workers in the field of Ascomycotina taxonomy. Unitunicate asci are bounded by a **single wall** which is thickened at the apex and is characteristically perforated by a pore through which ascospores are discharged simultaneously (see Section II, p. 171). Bitunicate asci, in contrast, are bounded by two distinct walls, the outer one of which ruptures during dehiscence, so enabling the elastic inner wall to expand to form a cylindrical sac. Ascospores are discharged successively through an elastic pore in the expanded inner sac.

The ascus in *Xylaria longpipes* Nitschke is both **struc-turally** and **functionally** unitunicate (figs. 287, 288. See also figs. 333–336). The wall consists of a single, homogeneous layer of granulo-fibrillar material, the fibrils being orientated circumferentially around the ascus (figs. 287, 288). Variation in fibril orientation occurs at the tip of the ascus where the **apical ring** joins the thickened wall (fig. 287).

The ascus of *Pleospora herbarum* (Fr.) Rabenh. is typically bitunicate. The outer wall (**exoascus**) is thin and inconspicuous, but the inner wall (**endoascus**) is thick, fibrillar and may be laminated (fig. 289). Fibrils towards the periphery of the endoascus run circumferentially, while those in the inner region assume conspicuous feather-like convolutions (fig. 289).

Additional reading

BECKETT, A. and CRAWFORD, R. M. (1973). The development and fine structure of the ascus apex and its role during spore discharge in *Xylaria longipes*. *New Phytol.*, **72**, 357.
REYNOLDS, D. R. (1971). Wall structure of a bitunicate ascus. *Planta (Berl.)*, **98**, 244.

Figs. 287, 288

Transverse sections through the unitunicate asci of *Xylaria longipes* Nitschke made in planes 1 and 2 as indicated in fig. 333. Note the single wall and arrangement of fibrils. Glutaraldehyde/acrolein–osmium tetroxide fixation. × 28,000.

Fig. 289

Transverse section of a bitunicate ascus of *Pleospora herbarum* (Fr.) Rabenh. Note the thin amorphous exoascus (arrows) and the laminated fibrillar endoascus. The dense, spherical structure within the ascus is an ascospore. Glutaraldehyde–osmium tetroxide fixation. × 8,400.

Figs. 287–9 *287, 288 From* BECKETT, A. and CRAWFORD, R. M. (1973). *New Phytol.*, **72**, 357. *289 Micrograph by* DR BRONWEN GRIFFITHS, Imperial College, London.

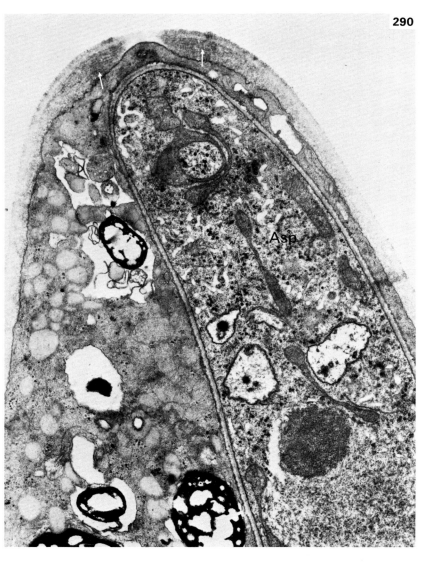

290

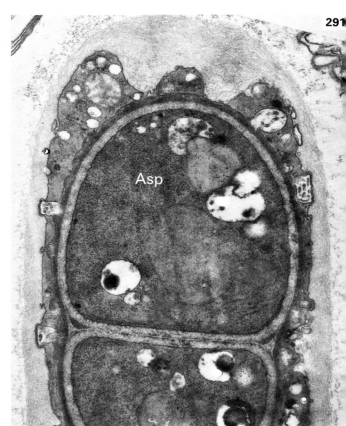

291

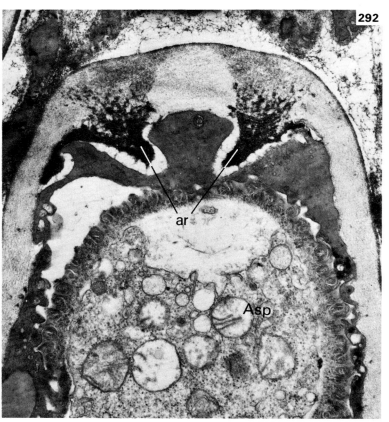

292

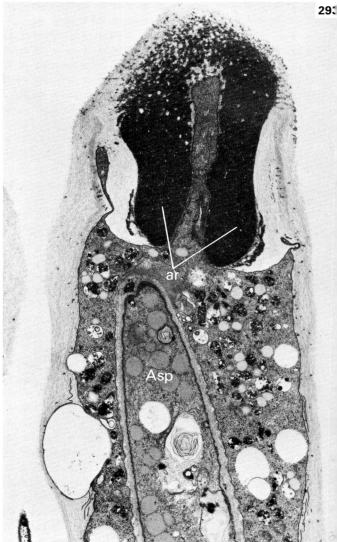

293

B. Reproductive Structures (cont.)

Sexual reproduction (cont.)

Ascus apex structure

Emphasis has recently been laid on the possible use and importance of the ascus apex structure as a taxonomic character and attempts have been made to base the taxonomy of some groups on this feature. It has been found that ascus apical structure is consistent within certain families though variable between them. For example, the inner wall layer at the tip of the ascus of *Lophodermella sulcigena* (Rostr.) v. Hohn. (Hypodermataceae) is locally differentiated to form an **annular thickening** around an electron-transparent, thin region, which is destined to become the **apical pore** (fig. 290).

In *Nectria episphaeria* (Tode ex Fr.) Fr. (Nectriaceae), the granulofibrillar ascus wall is thickened at the apex and the lower part of the thickening forms an annular projection into the ascus epiplasm (fig. 291). The wall of the ascus is a single layer and homogeneous throughout the thickening. In contrast to this the ascus apex of *Hypoxylon fuscum* (Pers. ex Fr.) Fr. (Xylariaceae) is somewhat more complex. The ascus wall which is thickened at the apex is granulo-fibrillar and has a thin electron-opaque surface layer (fig. 292). The lower part of the thickening forms a tapered annular projection into the epiplasm. Within the thickening additional materials are deposited to form a laminated electron-opaque **apical ring** continuous above with a moderately electron-opaque **apical cushion**. The base of the apical ring is extended outwards and downwards around the top of the epiplasm.

A similar but larger apical ring structure is found in the ascus of the related fungus *Rosellinia aquila* (Fr.) de Not. (Xylariaceae) (fig. 293). Here the apical ring extends into the ascus epiplasm for a distance of about $4 \cdot 0 – 5 \cdot 0$ μm enclosing a narrow channel which is continuous with the ascus lumen at the lower end and closed at the upper end by the thickened ascus wall and occluded base of the apical ring.

It has now been shown that ascus apical structures are intimately concerned with ascus dehiscence and spore discharge (see Section II, p. 171).

Additional reading

BECKETT, A. and CRAWFORD, R. M. (1973). The development and fine structure of the ascus apex and its role during spore discharge in *Xylaria longipes*. New Phytol., **72**, 357.

GRIFFITHS, H. BRONWEN (1973). Fine structure of seven unitunicate Pyrenomycete asci. *Trans. Br. mycol. Soc.*, **60**, 261.

Fig. 290

Section through the apical region of an ascus of *Lophodermella sulcigena* (Rostr.) v Hohn. The inner layer of the two-layered wall is locally thickened (arrows) around an electron-transparent apical pore region. Glutaraldehyde/formaldehyde–osmium tetroxide fixation. × 16,000.

Fig. 291

Section through the apical region of an ascus of *Nectria episphaeria* (Tode ex Fr.) Fr. showing the annular thickening projecting into the ascus lumen. Glutaraldehyde–osmium tetroxide fixation. × 21,800.

Fig. 292

The apical region of an ascus of *Hypoxylon fuscum* (Pers. ex Fr.) Fr. showing the annular thickening and the electron-opaque apical ring (ar). (Asp) ascospore. Glutaraldehyde–osmium tetroxide fixation. × 22,770.

Fig. 293

Section through the apical region of an ascus of *Rosellinia aquila* (Fr.) de Not. showing the conspicuous, electron-opaque apical ring (ar). Glutaraldehyde–osmium tetroxide fixation. × 10,200.

Figs. 290–3 *290 Micrograph by* DR R. CAMPBELL, Department of Botany, Bristol University. *291, 292 Micrographs by* DR BRONWEN GRIFFITHS, Imperial College, London. *293 Micrograph by* A. BECKETT, Bristol University.

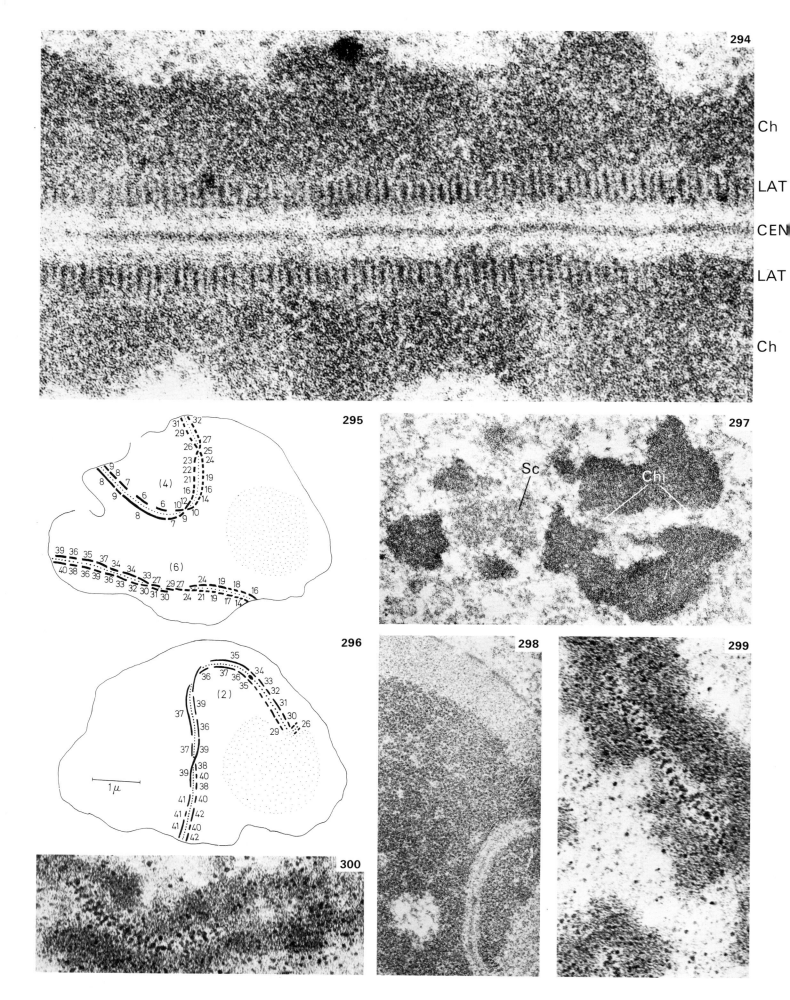

294

Ch

LAT

CEN

LAT

Ch

295

(4)

31 32
29 27
26 25
23 24
22 19
21 16
16 14
9 8 10 12
8 7 6 6 10
9 8 7 9 10

(6)

39 36 35 37 34 34 33
40 38 36 39 36 33 27 24 19 18 16
32 30 31 30 29 27 24 21 19 17 14

296

(2)

35
36 37 36 34
35 33
37 39 32
31
36 30
37 39 29 26
39 38
40
41 40
41 42
41 40
42

1μ

297

Sc

Chl

298

299

300

B. Reproductive Structures (cont.)

Sexual reproduction (cont.)

Meiosis: the synaptonemal complex and meiotic recombination

The cytological basis of meiotic recombination in Asco-mycotina at the light and electron microscopical level is most readily studied in *Neottiella rutilans* (Fr.) Dennis and *Neurospora crassa* Shear and Dodge. Sister chromatids of the unpaired **leptotene** chromosome are readily visible with the electron microscope (figs. 299, 300).

Pairing of the **homologous chromosomes** requires their rough alignment and thereafter the formation of the **synaptonemal complex**, a structure which holds the two homologous chromosomes of a bivalent in precise alignment along their entire length (figs. 294, 295, 296). In the split between the two sister chromatids of the un-paired chromosome a single **lateral component** is laid down. This component consists of **ribonucleoprotein** and can be seen labelled preferentially with silver grains (figs. 299, 300). The other part of the synaptonemal complex, the **central region**, appear first in the nucleolus which is always appressed to the nuclear envelope at this stage (fig. 298). For precise pairing the lateral component has to be relocated from its position between the sister chromatids into a position lateral to both sister chromatids. Thereby the chromatin of the two sister chromatids is brought into such a close physical association that they are no longer visible as separate entities. The lateral com-ponents of the homologous chromosomes are thereafter combined with the ribonucleoprotein of the central region into a single synaptonemal complex. At **pachytene** the synaptonemal complex (fig. 294) holds the homologous chromosomes closely paired at a distance of about 1,000 Å. It joins the homologous chromosomes over their entire length (figs. 295, 296) including **euchromatic** as well as **heterochromatic** regions and the homologous **centro-meres**. Reconstructions from serial sections have revealed the **telomeres** of all bivalents to be anchored to the nuclear envelope with the exception of chromosome arms which are attached to the nucleolus and carry the nucleolar organizer. The dimensions of the synaptonemal complex as shown here for ascomycetes are very similar in all eukaryotes of the plant and animal kingdoms that have so far been studied and are thus independent of chromosome size, morphology or DNA content. This observed evolu-tionary stability is intimately related to the universality of four strand crossing over at meiosis.

When pairing of the homologous chromosomes is ter-minated and replaced by their repulsion at early **diplotene**, the synaptonemal complex is shed from the **bivalents** except at places where the chromosomes are held together by **chiasmata** (fig. 297). At early diplotene the chiasmata (the visible result of crossing over) consist of short pieces of synaptonemal complexes. These short stretches are gradually modified in structure and eliminated prior to **diakinesis**, at which time the chiasmata consist of con-tinuous chromatin bridges, i.e. the broken and rejoined chromatids. The ring- and rod-shaped bivalents are strongly contracted and no longer attached with their telomeres at the nuclear envelope.

The synaptonemal complex in meiotic cells is the vector for chromosome pairing and crossing over as is shown by the genesis of the chiasmata from the synaptonemal complex.

Additional reading

GILLIES, C. B. (1972). Reconstruction of the *Neurospora crassa* pachytene karyotype from serial sections of synaptonemal complexes. *Chromo-soma (Berl.)*, **36**, 119.

WESTERGAARD, M. and WETTSTEIN, D. VON (1970). Studies on the mechanism of crossing over. IV. The molecular organization of the synapto-nemal complex in *Neottiella* (Cooke) Sac. (Ascomycetes). *C. r. Trav. Lab. Carlsberg*, **37**, 239.

WETTSTEIN, D. VON (1971). The synaptonemal complex and four strand crossing over. *Proc. natn. Acad. Sci. U.S.A.*, **68**, 851.

Fig. 294

Longitudinal section through a bivalent at pachytene of *Neottiella rutilans* (Fr.) Dennis. The two paired homologous chromosomes (Ch) are held in register by the synaptonemal complex which consists of two banded lateral components (LAT) separated by the 1,000 to 1,200 Å wide central region. Within the latter the central component (CEN) can be recognized. Glutaraldehyde–Dalton fixation. × 154,000.

Figs. 295, 296

Reconstructions of synaptonemal complexes from serial sections through a pachytene nucleus of *Neurospora crassa* Shear and Dodge. Numbers indicate sections in which portions of the synaptonemal complex occur and the dotted area represents the nucleolus. The bivalents of chromosomes 4 and 6 end with both telomeres at the nuclear envelope. Chromosome 2 carries the subterminal nucleolus organizer which serves as one anchoring site of the bivalent.

Fig. 297

Section through a bivalent at diplotene of *Neottiella rutilans*. The synaptonemal complex has been stripped from the bivalent and lies in disorganized form (Sc) between the repulsing homo-logues. At the site of chiasmata (Chi) small stretches of the synaptonemal complex have been retained, revealing that a chiasma originates from the synaptonemal complex. Glutar-aldehyde–osmium tetroxide fixation. × 47,000.

Figs. 298, 299, 300

Leptotene of *N. rutilans*. Aggregates of central regions of the synaptonemal complexes are found in the nucleolus prior to chromosome pairing (fig. 298). A single banded lateral com-ponent is formed in the split between the two sister chromatids of the unpaired chromosomes. In figs. 299 and 300 the banded lateral component has been labelled preferentially with silver grains. Fig. 298. Glutaraldehyde–Dalton fixation. × 41,000.

Figs. 294–300 *(299, 300 Formalin fixation–HCl. hydrolysis: stained with ammoniacal silver ions, pH 9·3. Both × 113,000.) 294, 298 From* WESTERGAARD, M. and WETTSTEIN, D. VON (1970). *C. r. Trav. Lab. Carlsberg*, **37**, 239. *295, 296 From* GILLIES, C. B. (1972). 'Reconstruction of the neurospora crassa pachytene karotype from serial sections of synaptonemal complexes', *Chromosoma*, (*Berl.*) **36**, 119–30. Berlin–Heidelberg–New York: Springer. *297, 300 From* WETTSTEIN, D. VON (1971). *Proc. natn. Acad. Sci. U.S.A.*, **68**, 851. *299 From* WESTERGAARD, M. and WETTSTEIN, D. VON (1972). 'The synaptinemal complex', *A. Rev. Genetics*, **6**, 71.

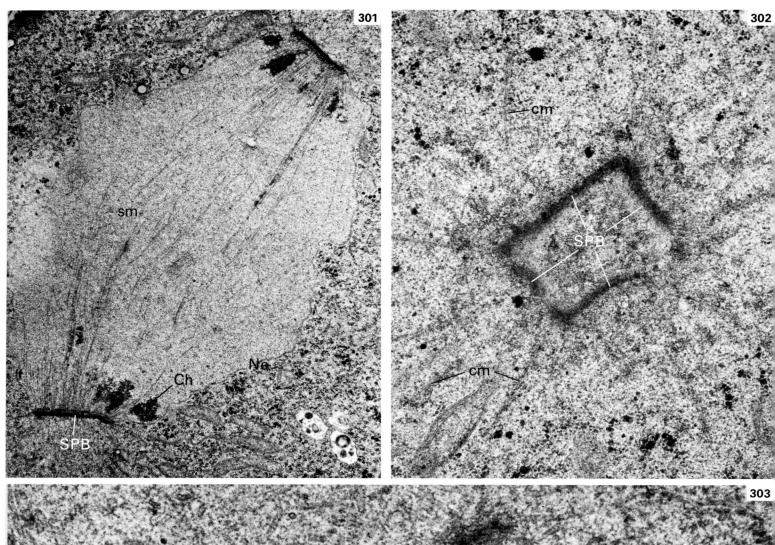

B. Reproductive Structures (cont.)

Sexual reproduction (cont.)

Meiosis: spindles and spindle pole bodies

The involvement of spindles and spindle pole bodies in the mechanics of chromosome separation has already been discussed with regard to mitosis (p. 93), and from a structural view point meiosis is fundamentally similar. However, it is with meiotic nuclei that some of the most detailed work has been done on the ultrastructure of spindle pole bodies and on their relationship with spindle microtubules.

In the Discomycete fungus *Ascobolus stercorarius* (Bull.) Schroet. the **intranuclear** spindle microtubules are associated, at each pole of the dividing nucleus, with an electron-opaque spindle pole body (fig. 301). In tangential section the spindle pole body is a rectangular, plate-like structure closely appressed to the outer surface of the nuclear envelope (figs. 302, 307a). Cytoplasmic microtubules (**astral ray microtubules**) radiate from the external surface of the spindle pole body (figs. 302, 303, 307a). At metaphase and anaphase of nuclear division, three zones are recognizable in the region of the spindle pole body. (1) An electron-opaque, **outer zone**, (2) a morphologically distinct region of the **nuclear envelope** and (3) an **inner zone**, which is somewhat amorphous and less dense and within which the microtubules of the spindle are apparently embedded (fig. 303).

The spindle is made up of two distinct sets of microtubules. Those which run from the spindle pole body and terminate within the chromosomes (**chromosomal microtubules**), and those which apparently extend throughout the length of the spindle and are not associated with chromosomes (**continuous microtubules**) (figs. 301, 303).

In the Pyrenomycete, *Podospora anserina* (Ces.) Niessl. the spindle pole body is associated at one end only with the nuclear envelope (figs. 304, 307b). It consists of two closely associated but rather irregular electron-opaque bands along the outer sides of which amorphous dense material is aggregated. The ends of cytoplasmic microtubules (**astral ray microtubules**) are embedded in this amorphous material (figs. 305, 307b).

We have seen that prior to **mitotic metaphase** spindle pole bodies undergo replication (Section II, p. 93). Similarly during the **pachytene** stage of **meiotic prophase**, spindle pole bodies divide, migrate around the nuclear envelope and subsequently occupy a position at the poles of the intranuclear spindle. At present little information is available on the process of spindle pole body replication but during pachytene of **post meiotic mitosis** in *Sordaria fimicola* (Roberge) Ces. and de Not. the organelle appears to split into two daughter halves which initially are connected with each other by a narrow electron-opaque band of material (fig. 306). This connecting structure is similar in form and position to the **plaque bridge**[1] seen connecting the daughter halves of yeast spindle pole bodies (Section II, p. 157), and to the so-called **middle piece** which links the globular ends of the spindle pole bodies in many Basidiomycotina (see Section III, p. 207, and fig. 405).

[1] The **plaque bridge** (MOENS, P. B. and RAPPORT, E. (1971). *J. Cell Biol.*, **50**, 344) may in fact be homologous with the **middle piece** of Basidiomycotina spindle pole bodies. However, the former term will be retained for reference to spindle pole bodies of *Saccharomyces cerevisiae* (see p. 157 and fig. 309).

Additional reading

BECKETT, A. and CRAWFORD, R. M. (1970). Nuclear behaviour and ascospore delimitation in *Xylosphaera polymorpha. J. gen. Microbiol.*, **63**, 269.
MOENS, P. B. and RAPPORT, E. (1971). Spindles, spindle plaques and meiosis in the yeast *Saccharomyces cerevisiae* Hansen. *J. Cell. Biol.*, **50**, 344.
ZICKLER, D. (1970). Division spindle and centrosomal plaques during mitosis and meiosis in some Ascomycetes. *Chromosoma (Berl.)*, **30**, 287.

Fig. 301

Longitudinal section through a nucleus of *Ascobolus stercorarius* (Bull) Schroet. at anaphase I. (i.e. division I of meiosis). Note relationships between the spindle pole bodies (SPB), the spindle microtubules (sm), chromosomes (Ch), and the intact nuclear envelope (Ne). Glutaraldehyde–osmium tetroxide fixation. × 14,000.

Fig. 302

A spindle pole body (SPB) sectioned tangentially and slightly outside the median plane, the nucleus being beneath the plane of section. Note cytoplasmic (astral ray) microtubules (cm). Glutaraldehyde–osmium tetroxide fixation. × 36,800.

Fig. 303

Section through the polar region of an intranuclear spindle at anaphase I. Associated with the spindle pole body (SPB) is an outer zone (oz), a modified region of nuclear envelope (Ne[1]) and an inner zone (iz). Continuous spindle microtubules (sm.c) and chromosomal microtubules (sm.ch) terminate at the inner zone. Cytoplasmic microtubules (cm) can be seen outside the nucleus. Note intact nuclear envelope (Ne). Glutaraldehyde–osmium tetroxide fixation. × 76,000.

Figs. 301–3 *301 From* ZICKLER, D. (1970). 'Division spindle and centrosomal plaques during mitosis and meiosis in some ascomycetes', *Chromosoma (Berl.)* **30**, 287–304. Berlin–Heidelberg–New York:Springer. *302 Micrograph by* DR DENISE ZICKLER, Laboratoire de Génétique, Orsay. *303 From* ZICKLER, D. (1969). *C. r. Acad. Sci. Paris*, **268**, 3040.

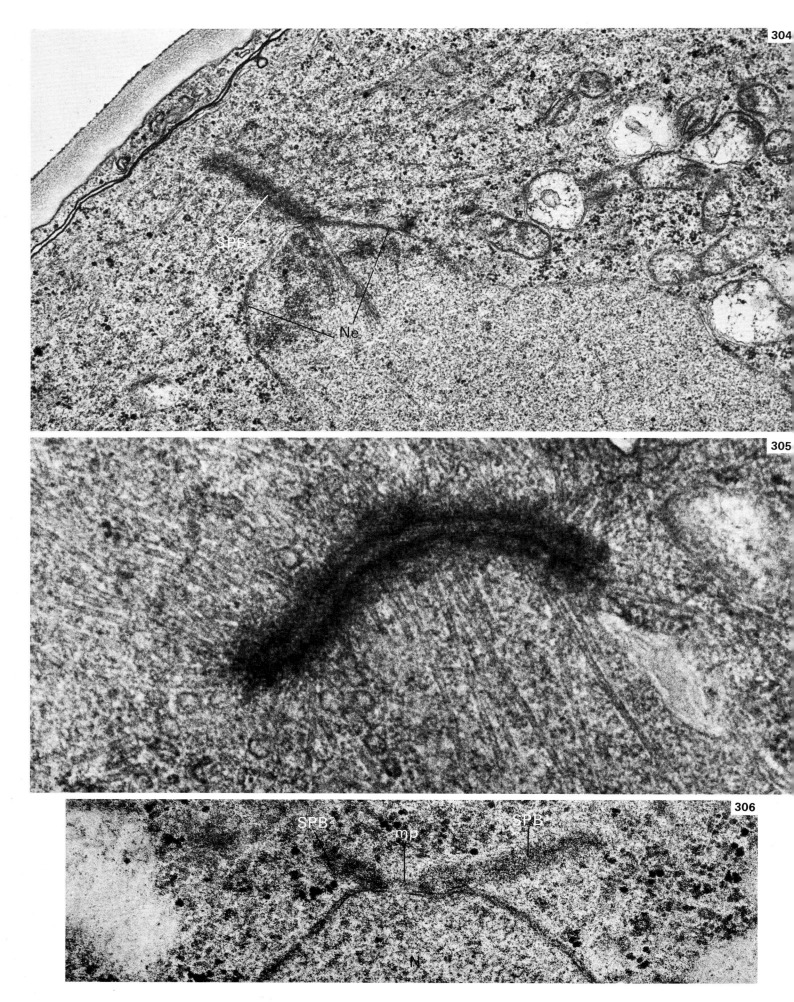

B. Reproductive Structures (cont.)
Sexual reproduction (cont.)
Meiosis: spindles and spindle pole bodies (cont.)

Fig. 304

Longitudinal section through the polar region of an anaphase III nucleus (i.e. post meiotic mitosis) of *Podospora anserina* (Ces.) Niessl. The association of one end of the spindle pole body (SPB) with the nuclear envelope (Ne) can be seen. Glutaraldehyde—osmium tetroxide fixation. × 33,600.

Fig. 305

A spindle pole body of *P. anserina* at high magnification. Note how the cytoplasmic (astral ray) microtubules terminate in electron-opaque material on either side of the spindle pole body. Glutaraldehyde—osmium tetroxide fixation. × 100,000.

Fig. 306

Part of a pachytene III nucleus (N) of *Sordaria fimicola* (Roberge) Ces. and de Not. The spindle pole body has divided into daughter halves (SPB[1] and SPB[2]). A narrow, electron-opaque middle piece (mp) connects the two. Glutaraldehyde—osmium tetroxide fixation. × 59,400.

Figs. 304—6 *304, 306 Micrographs by* DR DENISE ZICKLER, Laboratoire de Génétique, Orsay. *305 From* ZICKLER, D. (1970). *Abstr. 7th congrès M. Electronique, Grenoble,* **3**, 269.

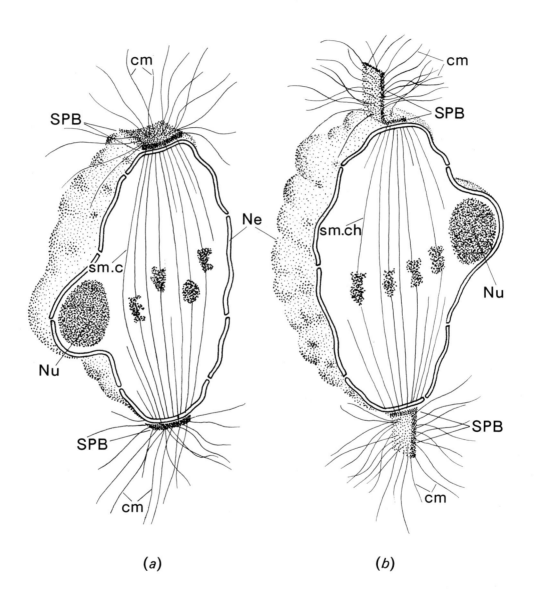

(a)   (b)

B. Reproductive Structures (cont.)

Sexual reproduction (cont.)

Meiosis: spindles and spindle pole bodies (cont.)

Fig. 307 (a), (b)

A three-dimensional representation of half nuclei of *Ascobolus* (a) and *Podospora* (b), illustrating the orientation and position of the spindle pole bodies (SPB) with respect to the nuclear envelope (Ne), chromosomal microtubules (sm.ch.), continuous microtubules (sm.c) cytoplasmic (astral ray) microtubules (cm) and the nucleolus (Nu).

Fig. 307 *Modified from* ZICKLER, D. (1970). 'Division spindle and centrosomal plaques during mitosis and meiosis in some ascomycetes', *Chromosoma*, **30**, 287–304, Berlin–Heidelberg–New York: Springer.

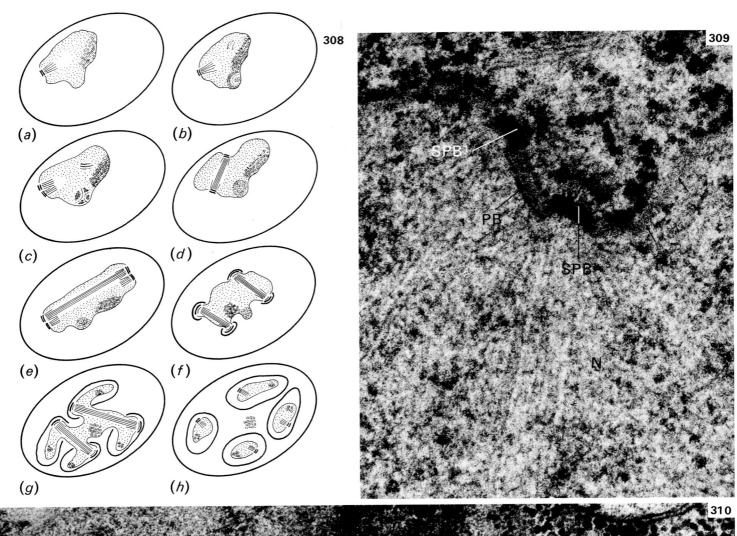

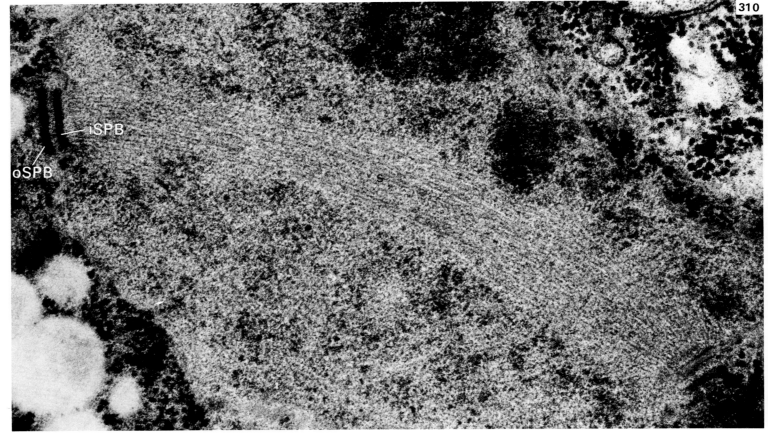

B. Reproductive Structures (cont.)
Sexual reproduction (cont.)
Ascospore delimitation

Early light microscopists coined the term **free cell formation** to describe the process of ascospore formation within an ascus. Electron microscopy has shown that in all Ascomycotina so far studied ascospore delimitation involves the compartmentalization of ascus cytoplasm by double membranes into nucleate portions. In the yeast *Saccharomyces cerevisiae* Hansen this process is intimately associated with the two meiotic divisions of the nucleus since it is from the polar regions of the meiotic spindles that the **ascospore-delimiting membranes**[1] are formed (figs. 308f–h, 312. See also fig. 273). Meiosis in *S. cerevisiae* is **uninuclear** in that both divisions occur within the one nucleus which is surrounded by an intact envelope (fig. 308a–g). The poles of the spindles are associated with **bipartite spindle pole bodies** (fig. 310), which divide into daughter halves. These are initially connected by a **plaque bridge** (fig. 309). During the second meiotic division, double, ascospore-delimiting membranes form in the region external to the outer spindle pole body (figs. 308f, 312). Progressive envelopment by these membranes, of the lobes of the dividing nucleus, results in the formation of four **haploid, uninucleate ascospores** (fig. 308g, h, see also fig. 274).

In *Taphrina deformans* (Berk.) Tul. ascospore-delimiting membranes are formed as extensions of the ascus **plasma membrane** which invaginates at **specific points** adjacent to each nucleus after the post meiotic mitosis is complete (figs. 313–316). The nuclei usually lie close to the ascus wall and continuities between the double delimiting mem-

[1] The term **'ascospore-delimiting membranes'** is used here to denote the two unit membranes which envelope the nuclei during ascospore formation in all Ascomycotina. These membranes are equivalent to the **'prospore wall'** [MOENS, P. B. (1971). *Can. J. Microbiol.*, **17**, 507.] and to the **'forespore membranes'** [GUTH, E., HASHIMOTO, T. and CONTI, S. F. (1972). *J. Bact.*, **109**, 869.]

branes and the plasma membrane can be seen at these points (fig. 313). A **spindle pole body** is associated with the interphase nucleus at the time of its envelopment by the delimiting membranes (fig. 314). When treated with a phosphotungstic acid-chromic acid mixture, which is a preferential stain for plasma membrane, we find that the ascus plasma membrane and the spore-delimiting membranes are the only membranes of the cell to show a positive staining reaction (figs. 315, 316). These observations are interpreted as evidence for a **similarity** between the two membrane types, for a **difference** between them and the other cell membranes and for the **origin** of spore delimiting membranes at the plasma membrane. In certain Pyrenomycete and Discomycete fungi the double spore-delimiting membranes (fig. 317) lie around the ascus periphery and form a more or less complete sac surrounding the eight nuclei which are the products of the three nuclear divisions within the ascus. The origin of this membrane sac or **ascus vesicle** is at present uncertain although double membrane-bound vesicles have been observed forming by budding of the nuclear envelope (fig. 319). Membrane may be continuously added to the ascus vesicle by the fusion of smaller vesicles with it (fig. 318). The factors controlling the invagination of the ascus vesicle around individual nuclei are unknown but it has been suggested that the **cytoplasmic microtubules** (astral ray microtubules) which radiate from the spindle pole body, which is associated with the interphase nucleus (fig. 321), may act as a template for the enveloping membrane and/or as contractile elements which might physically 'pull in' the membranes around each nucleus (figs. 320, 321, 322). Once delimitation is complete, the inner of the ascospore-delimiting membranes forms the spore **plasma membrane**, while the outer one forms the spore **investing membrane**.

Additional reading

BECKETT, A. and CRAWFORD, R. M. (1970). Nuclear behaviour and ascospore delimitation in *Xylosphaera polymorpha. J. gen. Microbiol.*, **63**, 269.

MOENS, P. B. (1971). Fine structure of ascospore development in the yeast *Saccharomyces cerevisiae. Can. J. Microbiol.*, **17**, 507.

MOENS, P. B. and RAPPORT, E. (1971). Spindles, spindle plaques, and meiosis in the yeast *Saccharomyces cerevisiae* Hansen. *J. Cell. Biol.*, **50**, 344.

SYROP, M. J. and BECKETT, A. (1972). The origin of ascospore-delimiting membranes in *Taphrina deformans. Arch. Mikrobiol.*, **86**, 185.

Fig. 308 (a)–(h)
Diagrammatic summary of the processes of meiosis and ascospore delimitation in *Saccharomyces cerevisiae* Hansen.

Fig. 309
Part of a meiotic prophase nucleus (N) of *S. cerevisiae* showing the replicated spindle pole bodies (SPB[1] and SPB[2]) lying side by side and connected to each other by an amorphous band of material, the plaque bridge (PB). Note the nuclear envelope (arrows). Glutaraldehyde–osmium tetroxide fixation. × 140,000.

Fig. 310
An elongated intranuclear spindle (s) of *S. cerevisiae* showing inner (iSPB) and outer (oSPB) spindle pole bodies at either end. Glutaraldehyde–osmium tetroxide fixation. × 100,000.

Figs. 308–10 *308 By* DR P. B. MOENS, Department of Biology, York University, Ontario. *309 From* MOENS, P. B. and RAPPORT, E. (1971). *J. Cell Biol.*, **50**, 344. *310 Micrograph by* DR P. B. MOENS, Department of Biology, York University, Ontario.

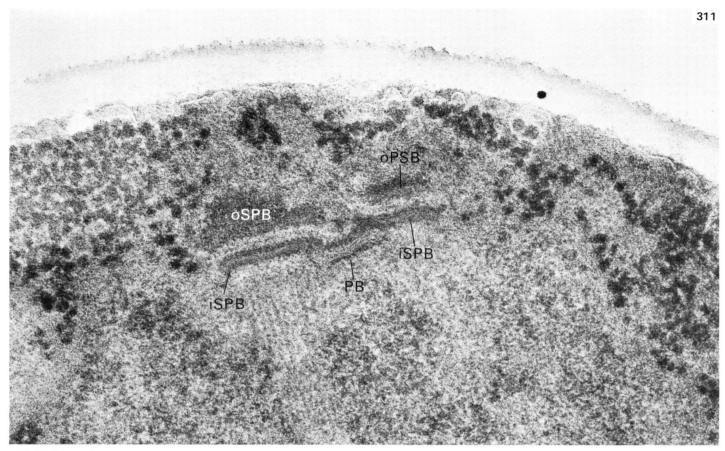

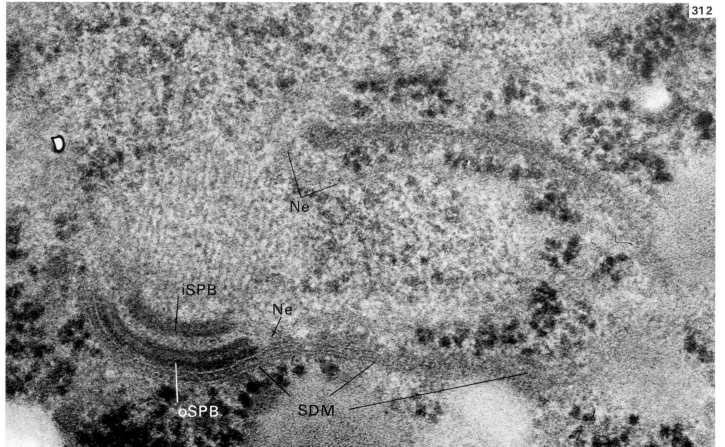

B. Reproductive Structures (cont.)
Sexual reproduction (cont.)
Ascospore delimitation (cont.)

Fig. 311

Spindle pole body replication prior to spindle formation and initiation of the second meiotic division. Inner (iSPB) and outer (oSPB) spindle pole bodies can be seen in each daughter pair. A plaque bridge (PB) links the daughter pairs of spindle pole bodies. Glutaraldehyde—osmium tetroxide fixation. ×140,000.

Fig. 312

Part of a nucleus of *S. cerevisiae* at meiosis II showing an early stage in the formation of ascospore-delimiting membranes (SDM). (iSPB) inner spindle pole body; (oSPB) outer spindle pole body; (Ne) nuclear envelope. Glutaraldehyde—osmium tetroxide fixation. ×145,000.

Figs. 311–12 *311 Micrograph by* DR P. B. MOENS, Department of Biology, York University, Ontario. *312 From* MOENS, P. B. and RAPPORT, E. (1971). *J. Cell Biol.,* **50**, 344.

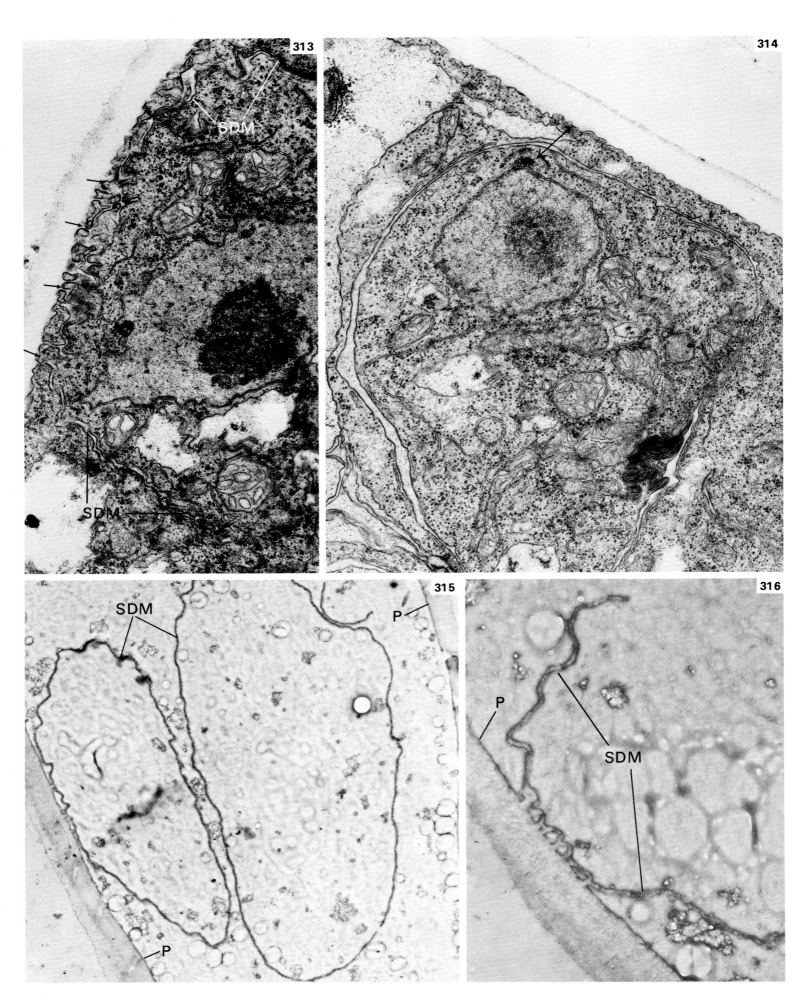

B. Reproductive Structures (cont.)

Sexual reproduction (cont.)

Ascospore delimitation (cont.)

Fig. 313

Part of an ascus of *Taphrina deformans* (Berk.) Tul. showing structural continuity (arrows) between the ascus plasma membrane and the spore-delimiting membranes (SDM). × 31,500.

Fig. 314

Part of an ascus of *T. deformans* showing a young spore initial containing an interphase nucleus with a spindle pole body (arrow). × 29,500.

Figs. 315, 316

Parts of asci of *T. deformans* stained with phosphotungstic acid–chromic acid showing the positive reaction with the ascus plasma membrane (P) and the spore delimiting membranes (SDM). Note continuity between the two membrane systems (fig. 316). Potassium permanganate fixation. × 8,850; × 21,750.

Figs. 313–16 (*313, 314 Glutaraldehyde/formaldehyde–osmium tetroxide fixation.*) *Micrographs by* MARY SYROP, Department of Botany, Bristol University. *315, 316 From* SYROP, M. J. and BECKETT, A. (1972). *Arch. Mikrobiol.,* **86**, 185.

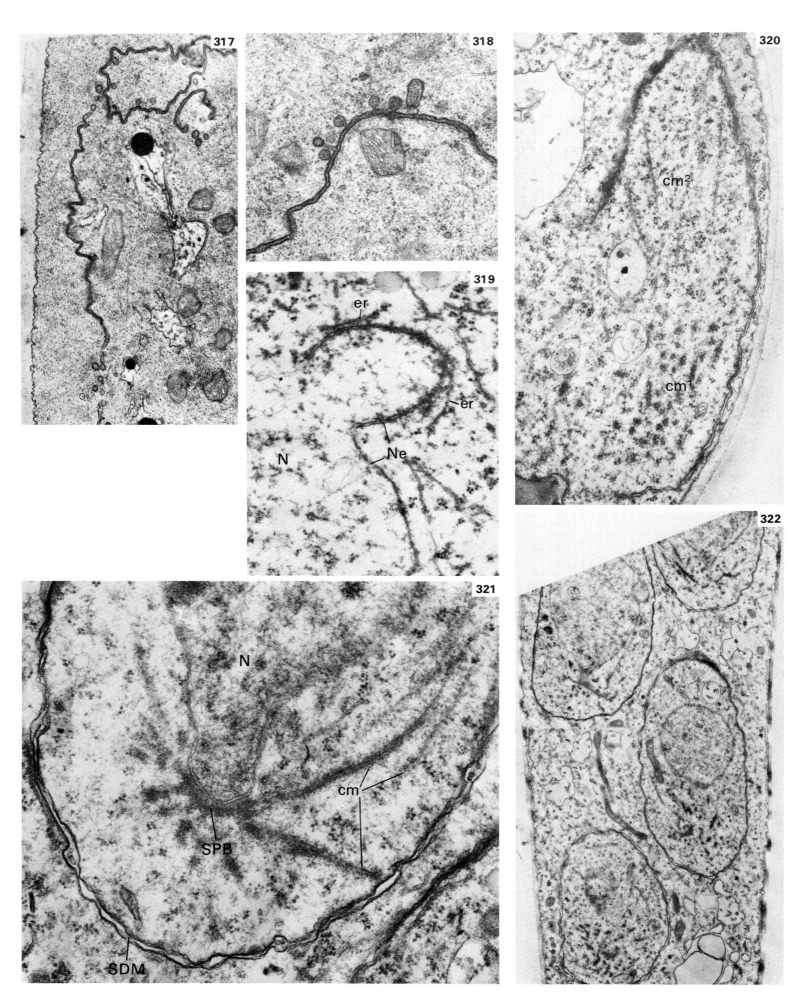

B. Reproductive Structures (cont.)
Sexual reproduction (cont).
Ascospore delimitation (cont.)

Figs. 317, 318

Parts of asci of *Lophodermella sulcigena* (Rostr.) v Hohn. showing the ascus vesicle which lines the ascus periphery. Note fusion of small vesicles with ascus vesicle membranes (fig. 318). Glutaraldehyde/formaldehyde—osmium tetroxide fixation. × 22,000; × 33,000.

Fig. 319

Part of a meiotic nucleus (N) of *Xylaria polymorpha* (Pers.) Grev. showing budding of the nuclear envelope (Ne) in association with endoplasmic reticulum (er). × 40,000.

Fig. 320

Part of an ascus of *X. polymorpha* showing an early stage in the invagination of the ascus vesicle which is associated with cytoplasmic microtubules. These microtubules are sectioned both transversely (cm¹) and longitudinally (cm²). × 25,000.

Fig. 321

Longitudinal section through part of an ascospore initial of *X. polymorpha* showing the nucleus (N), spindle pole body (SPB), cytoplasmic microtubules (cm) and spore-delimiting membranes (SDM). × 40,000.

Fig. 322

Longitudinal section through part of an ascus of *X. polymorpha* showing parts of four young spores. × 9,000.

Figs. 317–22 (*319–22 Glutaraldehyde/acrolein—osmium tetroxide fixation.*) *317 Micrograph by* DR R. CAMPBELL, Department of Botany, Bristol University. *318 From* CAMPBELL, R. (1973) *Protoplasma,* **78**, 69. *319–22 From* BECKETT, A. and CRAWFORD, R. M. (1970). *J. gen. Microbiol.,* **63**, 269.

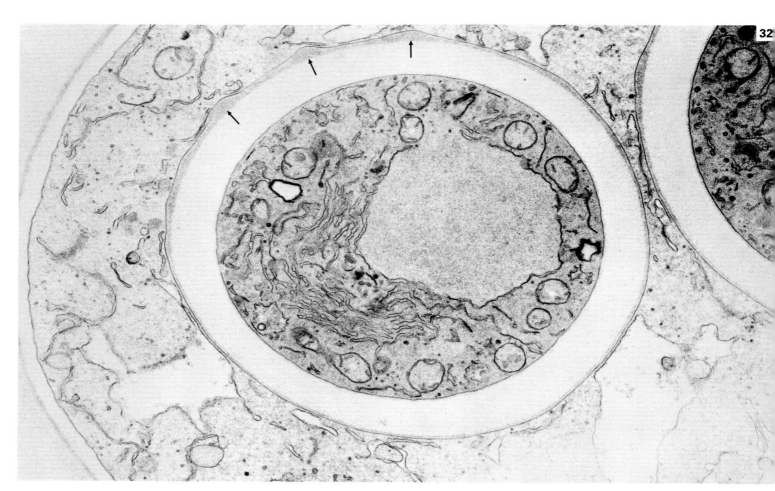

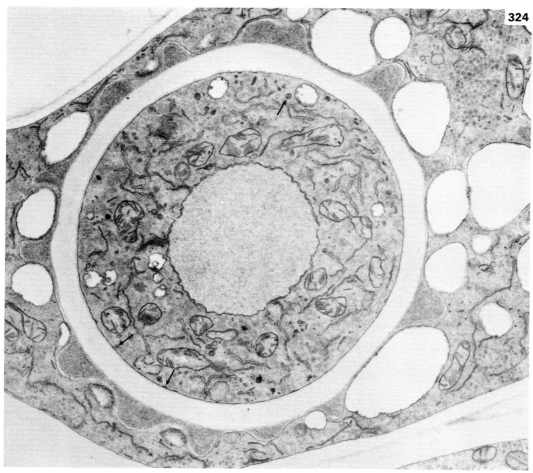

B. Reproductive Structures (cont.)
Sexual reproduction (cont.)
Ascospore maturation

Wall formation in ascospores is unique in that it occurs **within** the spore-producing cell (ascus) and **between** the spore plasma membrane and spore investing membrane. The process of primary wall deposition is similar for all Ascomycotina so far examined, but secondary and tertiary wall formation may follow varied patterns.

Mature ascospores of *Ascodesmis sphaerospora* (Seav.) Obrist typically have a reticulate outer wall. In young spores the reticulum develops as an electron-opaque layer to the outside of an electron-transparent **primary wall** (figs. 323, 324). Early in the sequence of wall deposition the reticulate pattern is evident (fig. 323). As the wall sculpturing becomes more pronounced, the vacuolar system in the **epiplasm** becomes divided into numerous small vacuoles which appear to fit into the pockets of the reticulum on each spore (fig. 324, 326). A morphogenetic role for the vacuolar system in determining the pattern of reticulation is, however, unlikely since vacuolar division does not occur until a late stage in development. As the reticulum develops, small electron-dense vesicles arise in the spore cytoplasm and appear to fuse with the plasma membrane, depositing their contents into the matrix of the primary wall (figs. 324, 326). In nearly mature ascospores a convoluted, thread-like substructure becomes evident within the matrix of the **secondary wall**, and an **intercalary layer** appears between the primary and secondary wall (fig. 326).

In ascospores of the mutant 'bumpy' strain the reticulate deposition of secondary wall material is less pronounced and moreover is hidden by the simultaneous deposition of wall material with staining properties similar to those of the primary wall (figs. 327, 328). During the final stages of spore maturation in this strain, large, lightly stained vesicles appear in the spore cytoplasm, suggesting that the wall material originates from within the spore, not in the epiplasm.

Additional reading

MAINWARING, H. R. (1972). The fine structure of ascospore wall formation in *Sordaria fimicola. Arch. Mikrobiol.*, **81**, 126.
WELLS, K. (1972). Light and electron microscopic studies of *Ascobolus stercorarius* II. Ascus and ascospore ontogeny. *University of California Publications in Botany*, **62**, 1. University of California Press.

Fig. 323

A very young ascospore of *Ascodesmis sphaerospora* (Seav.) Obrist. Secondary wall deposition has just begun. Beginnings of reticulate sculpturing are already evident (arrows). × 21,000.

Fig. 324

Young ascospore. Reticulate deposition of outer wall is more evident. Note small electron-dense inclusions in the cytoplasm of the spore and adjacent to the plasma membrane (arrows). × 14,000.

Figs. 323–4 *Potassium permanganate fixation. Micrographs by* DR G. C. CARROLL, Department of Biology, University of Oregon, Eugene, Oregon.

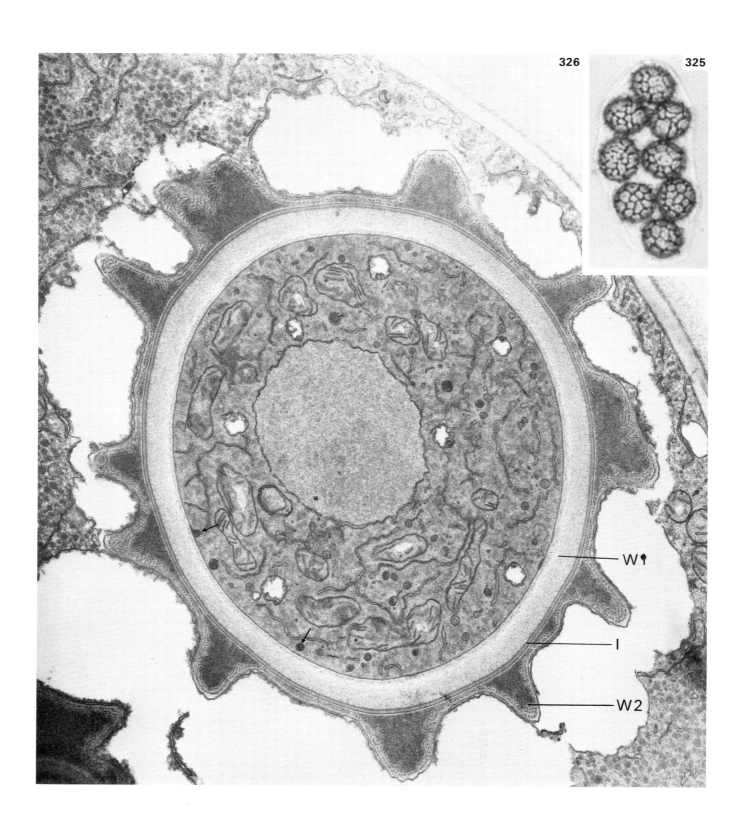

326

325

W♀

I

W2

B. Reproductive Structures (cont.)
Sexual reproduction (cont.)
Ascospore maturation (cont.)

Fig. 325

Light micrograph of mature ascus with eight reticulate ascospores.
×1,000.

Fig. 326

Electron micrograph of a nearly mature reticulate ascospore

showing the primary and secondary wall layers (W1, 2) and the intercalary layer (I). Note electron-dense bodies fusing with plasma membrane (arrows). *Potassium permanganate fixation.* ×21,000.

Figs. 325–6 *Micrographs by* DR G. C. CARROLL, Department of Biology, University of Oregon, Eugene, Oregon.

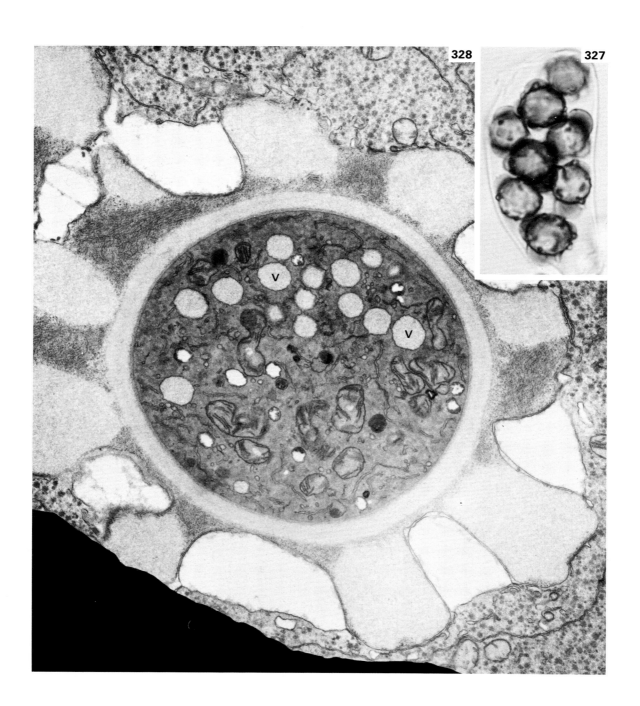

B. Reproductive Structures (cont.)

Sexual reproduction (cont.)

Ascospore maturation (cont.)

Fig. 327

Light micrograph of ascus with eight mutant 'bumpy' ascospores. × 1,000.

Fig. 328

Electron micrograph of mutant 'bumpy' ascospore. The large vesicles (v) in the spore cytoplasm show the same electron-staining properties as the material of which the warts on the outer wall are composed. *Potassium permanganate fixation.* × 21,000.

Figs. 327–8 *Micrographs by* DR G. C. CARROLL, Department of Biology, University of Oregon, Eugene, Oregon.

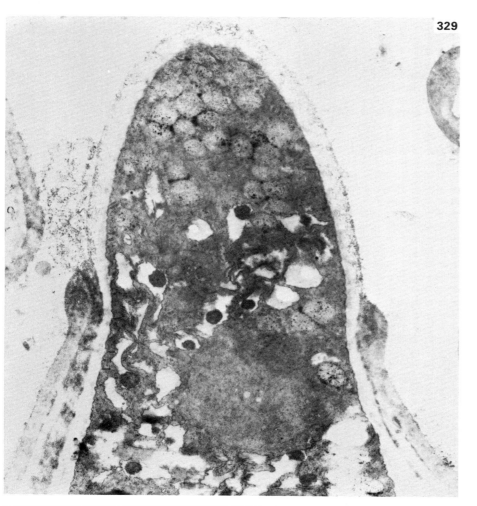

329

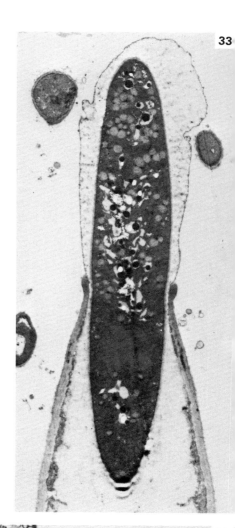

33

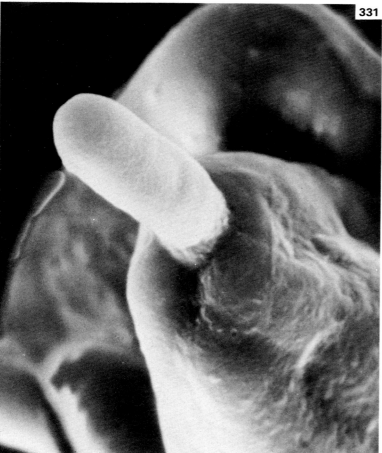

331

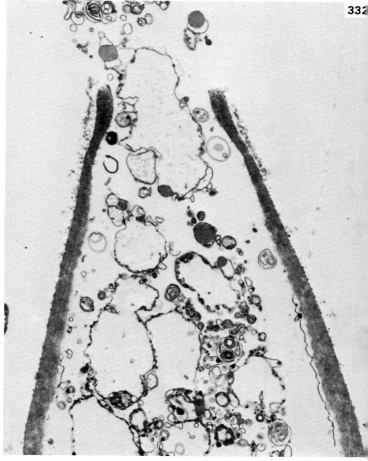

332

B. Reproductive Structures (cont.)
Sexual reproduction (cont.)
Ascospore discharge

The mode of ascus dehiscence and therefore also the method of spore discharge varies according to whether the ascus is **unitunicate** or **bitunicate** (see also p. 145). In the unitunicate ascus of *Lophodermella sulcigena* (Rostr.) v Hohn. the clavate ascospores are discharged successively through a pore in the expandable wall at the ascus tip (figs. 329, 330, 331). Each spore is held by the elastic wall surrounding the pore.

A high **osmotic pressure** is assumed to provide the motive force for active discharge of spores. Surrounding each spore is a mucilaginous sheath which swells owing to absorption of water prior to spore discharge. Pressure generated by water absorption may be sufficient to shoot the spores out. Once the major part of the spore has passed through the pore, the squeezing of the ascus wall against the tapering end of the spore (fig. 330) aids in throwing the spore clear of the ascus. With the liberation of one spore,

the next in line is forced into the pore, thus sealing it and retaining the internal pressure. When all eight spores have been discharged the vacuolated epiplasm is lost through the open pore (fig. 332).

In *Xylaria longipes* Nitschke the elaborate apical apparatus of the unitunicate ascus is involved in ascospore discharge. Maturing ascospores are normally positioned in a **uniseriate** arrangement beneath the apical ring. The apical spore is attached to the inside rim of the ring (and to the other spores behind it) by a band of fibrous material (figs. 333, 334). This linking of spores together in a 'chain' and to the apical ring may serve to align them beneath the pore which forms as a result of either mechanical rupture or enzymic-dissolution of the ascus wall at the proximal end of the apical ring. With the release of pressure the entire apical ring everts (fig. 336) and in doing so could jerk out the spores simultaneously.

Additional reading

BECKETT, A. and CRAWFORD, R. M. (1973). The development and fine structure of the ascus apex and its role during spore discharge in *Xylaria longipes*. *New Phytol.*, **72**, 357.
CAMPBELL, R. (1973). Ultrastructure of asci, ascospores and spore release in *Lophodermella sulcigena* (Rostr.) v Hohn. *Protoplasma*, **78**, 69.
FURTADO, J. S. and OLIVE, L. S. (1970). Ascospore discharge and ultrastructure of the ascus in *Leptosphaerulina australis*. *Nova Hedwigia*, **19**, 799.

Fig. 329

Longitudinal section through the tip of a dehiscing ascus of *Lophodermella sulcigena* (Rostr.) v Hohn. from which an ascospore is just emerging. Note how the spore is constricted at the point where it passes through the apical pore. Glutaraldehyde/formaldehyde—osmium tetroxide fixation. × 17,000.

Fig. 330

Section through part of an ascus of *L. sulcigena* showing a spore which has passed through the pore to midway along its length. The extruded part of the spore is covered by a mucilaginous sheath. Glutaraldehyde/formaldehyde—osmium tetroxide fixation. × 5,000.

Fig. 331

Stereoscan electron micrograph of a dehiscing ascus at a similar stage to that shown in fig. 330. × 9,000.

Fig. 332

Section through the tip of a dehisced ascus. Glutaraldehyde/formaldehyde—osmium tetroxide fixation. × 9,000.

Figs. 329–32 *329, 332 Micrographs by* DR R. CAMPBELL, Department of Botany, Bristol University. *330, 331 From* CAMPBELL, R. (1973). *Protoplasma*, **78**, 69.

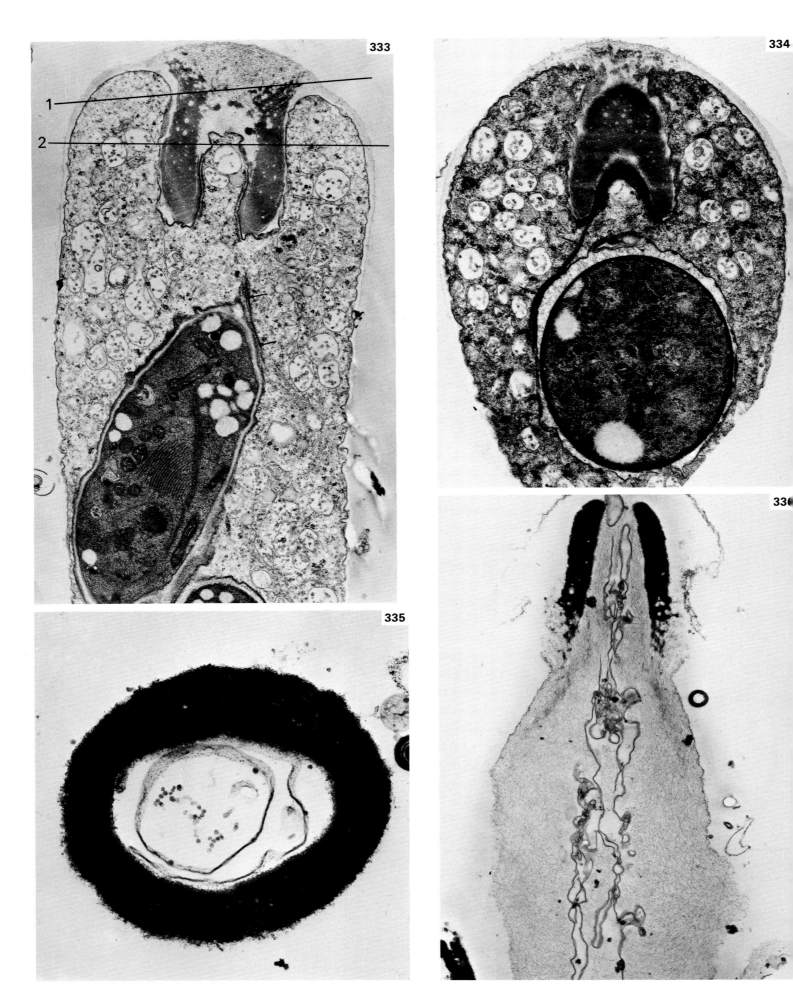

B. Reproductive Structures (cont.)
Sexual reproduction (cont.)
Ascospore discharge (cont.)

Fig. 333

Longitudinal section through part of an ascus of *Xylaria longipes*
Nitschke showing the position of the apical spore beneath the
apical ring. Note fibrous material at one end of spore (arrows).
× 17,400.

Fig. 334

Oblique longitudinal section through the apical region of an
ascus of *X. longipes* showing the fibrous material connecting the
spore with the inside of the apical ring (arrows). × 20,000.

Fig. 335

Transverse section of a detached, everted apical ring. × 40,000.

Fig. 336

Longitudinal section through a dehisced ascus. Note how the
apical ring has everted. Compare with fig. 333. × 11,600.

Figs. 333–6 *Glutaraldehyde/acrolein–osmium tetroxide fixation.
From* BECKETT, A. and CRAWFORD, R. M. (1973). *New Phytol.,*
72, 357.

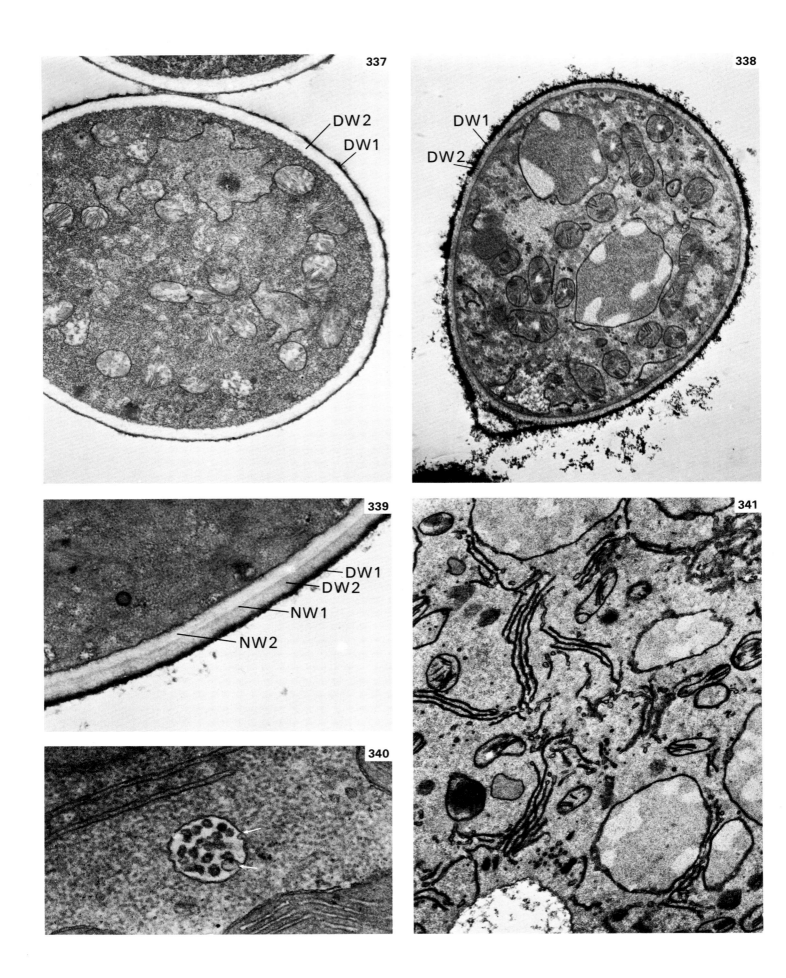

B. Reproductive Structures (cont.)
Spore germination
Conidia

Prior to germination in Deuteromycotina and Ascomycotina a number of changes in the ultrastructural morphology of the dormant spore take place which culminate in the emergence of one or more **germ tubes**. These changes, which include an increase in numbers of organelles such as mitochondria, vesicles, vacuoles and nuclei; an increase in endoplasmic reticulum, and a synthesis of one or more new wall layers, are characteristic of both asexual spores (**conidia**) and sexual spores (**ascospores**).

Ungerminated conidia of *Botrytis cinerea* Pers. possess a two-layered wall and contain several mitochondria and up to six nuclei (figs. 337, 338). Soon after inoculation on to a suitable nutrient medium the conidia swell, and two new wall layers are formed (fig. 339), to the inside of the original wall of the spore. Within the protoplast of the germinating spore a complex system of endoplasmic reticulum is formed, together with vacuoles, storage bodies and an increase in numbers of mitochondria (figs. 341–344). The emergence of the germ tube follows the synthesis of a third new wall layer (figs. 342, 343) which extends out and around the growing germ tube. This third wall layer is synthesized at a fixed locus surrounding the point of germ tube emergence, a process reminiscent of that which occurs during **secondary conidium** formation at the tip of a phialide (section II, p. 117).

Soon after the germ tube has grown out a perforate septum forms at the point where the germ tube leaves the conidium (figs. 345, 346, 344). **Multivesicular bodies** are frequently found in growing germ tubes (fig. 340).

Additional reading

GULL, K. and TRINCI, A. P. J. (1971). Fine structure of spore germination in *Botrytis cinerea*. *J. gen. Microbiol*, **68**, 207.

Figs. 337, 338

Ungerminated conidia of *Botrytis cinerea* Pers. showing the two layers of the wall (DW1, DW2). Kellenberger's osmium tetroxide, and potassium permanganate fixation respectively. ×17,330; ×14,210.

Fig. 339

Part of a germinating conidium showing the new wall layers (NW1, NW2) within the original wall layers (DW1, DW2). Glutaraldehyde–osmium tetroxide fixation. ×32,000.

Fig. 340

Part of a germ tube of *B. cinerea* showing a multivesicular body. Note invaginations in the bounding membrane (arrows). Glutaraldehyde–osmium tetroxide fixation. ×72,000.

Fig. 341

Part of the cytoplasm of a germinating conidium of *B. cinerea* showing the abundance of endoplasmic reticulum. Potassium permanganate fixation. ×17,800.

Figs. 337–41 *337–40 From* GULL, K. and TRINCI, A. P. J. (1971). *J. gen. Microbiol.*, **68**, 207. *341 Micrograph by* DR K. GULL, Queen Elizabeth College, London.

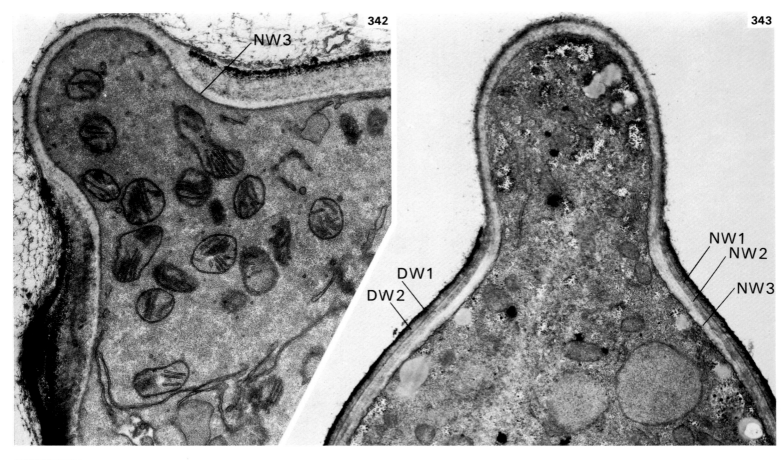

342

343

NW3

DW1
DW2
NW1
NW2
NW3

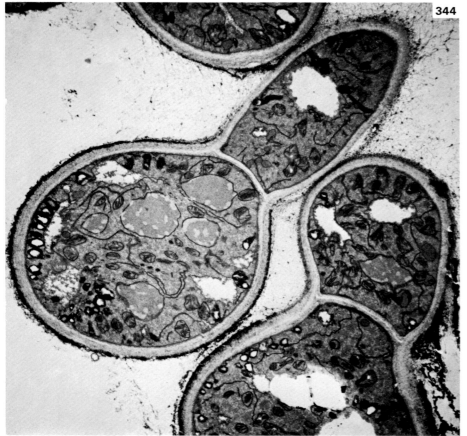

344

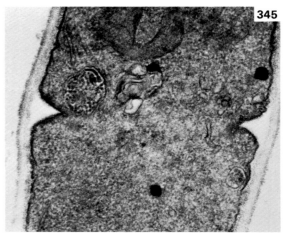

345

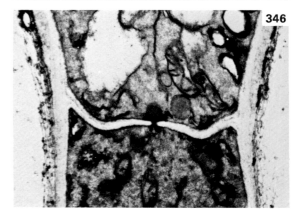

346

B. Reproductive Structures (cont.)
Spore germination (cont.)
Conidia (cont.)

Fig. 342

Median section through an emerging germ tube of *B. cinerea* showing the new wall layer (NW3) which forms the germ tube wall. Potassium permanganate fixation. × 23,400.

Fig. 343

A newly emerged germ tube showing the wall layers (DW1, DW2, NW1, NW2, NW3). Glutaraldehyde–osmium tetroxide fixation. × 14,900.

Fig. 344

Two conidia with germ tubes. Septa have formed at the base of the germ tubes. Potassium permanganate fixation. × 6,800.

Fig. 345

A developing septum showing the triangular septal initials as seen in transverse section. Glutaraldehyde–osmium tetroxide fixation. × 39,600.

Fig. 346

A fully developed septum showing the central pore and associated structures. Potassium permanganate fixation. × 17,000.

Figs. 342–6 *From* GULL, K. and TRINCI, A. P. J. (1971). *J. gen. Microbiol.,* **68**, 207.

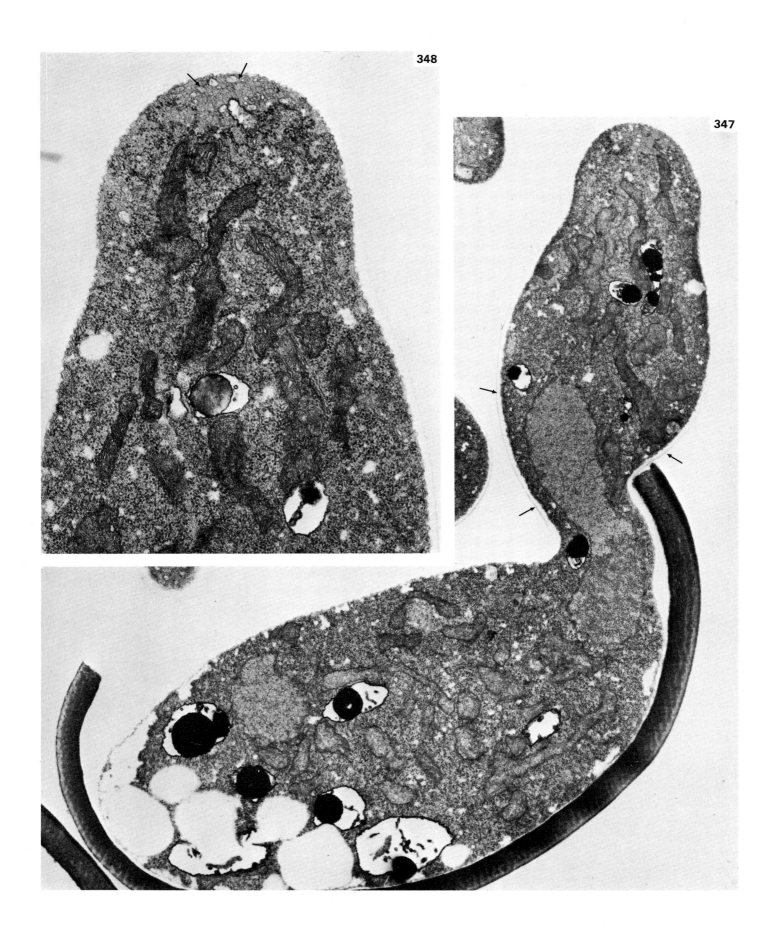

B. Reproductive Structures (cont.)
Spore germination (cont.)
Ascospores

When placed on a suitable nutrient medium, dormant ascospores of *Daldinia concentrica* (Bolton ex Fr.) Ces. and de Not. increase in size due to **spherical growth** as a result of which the thin outer layer of the three-layered spore wall splits and the spore is released from within it. The thick, melanized layer of the two remaining wall layers (fig. 347) is partially perforated by a narrow germ-fissure which runs along the length of the spore. During spherical growth this fissure gapes open to expose the inner layer of the wall which encloses the spore protoplast (fig. 347).

Prior to the emergence of the germ tube, wall vesicles aggregate at one point beneath the inner wall layer and **polarized growth** results in the formation of a slightly inflated germ tube which subsequently tapers at the tip and grows apically to produce a hyphal filament (fig. 347). Vesicles are present in a ribosome-free **zone of exclusion** at the tip of the growing germ tube (fig. 348; cf. fig. 93 and hyphal tips, figs. 1, 109, 145, 146, 350).

Additional reading

LOWRY, R. J. and SUSSMAN, A. S. (1968). Ultrastructural changes during germination of ascospores of *Neurospora tetrasperma*. *J. gen. Microbiol.*, **51**, 403.

Fig. 347
Transverse section through a germinating ascospore of *Daldinia concentrica* (Bolton ex Fr.) Ces. and de Not. showing the emerging germ tube (in longitudinal section) surrounded by an electron-transparent wall layer (arrows) which is formed as a new layer inside the inner wall layer of the spore. The thick, melanized middle wall layer of the spore gapes open along the germ-fissure. The outer wall layer has been lost during spherical growth. × 24,000.

Fig. 348
Median longitudinal section through the tip of a germ tube of *D. concentrica* showing the aggregation of vesicles (arrows) within an amorphous zone of exclusion. × 39,000.

Figs. 347, 348 *Glutaraldehyde/formaldehyde–osmium tetroxide fixation. Micrographs by* A. BECKETT, *Bristol University.*

SECTION 3

Basidiomycotina

Introduction

Ultrastructural evidence, in addition to morphological and biochemical evidence, supports the concept that Basidiomycotina represent a distinct group of fungi with evolutionary affinities with Ascomycotina. An apical body or **Spitzenkörper** is present during hyphal tip growth in Basidiomycotina. In the rusts the hyphae contain septa similar to those of Ascomycotina; however, most Basidiomycotina have an elaborate **septal pore apparatus** not encountered in other fungi. The Golgi apparatus of Basidiomycotina consists of individual **Golgi cisternae** not organized into stacks and is similar to that of Zygomycotina and Ascomycotina.

The manner of nuclear division in Basidiomycotina is distinctive. The **spindle pole body** before division usually consists of a **bipolar structure** which is also encountered in Ascomycotina. However, unlike that in Ascomycotina, the spindle pole body enters the nucleus at division. The breakdown of the nuclear envelope at division is characteristic of Basidiomycotina and has been found in the Zygomycotina.

The process of basidiospore formation appears to be unique. Basidiospores arise **exogenously**, their outer wall appearing to form from a thin layer continuous with the basidial wall. The main walls of the basidiospore are deposited outside the spore plasma membrane. In contrast, ascospore walls develop within the ascus either within an invagination of its plasma membrane or in an ascus vesicle (see Section II, p. 165).

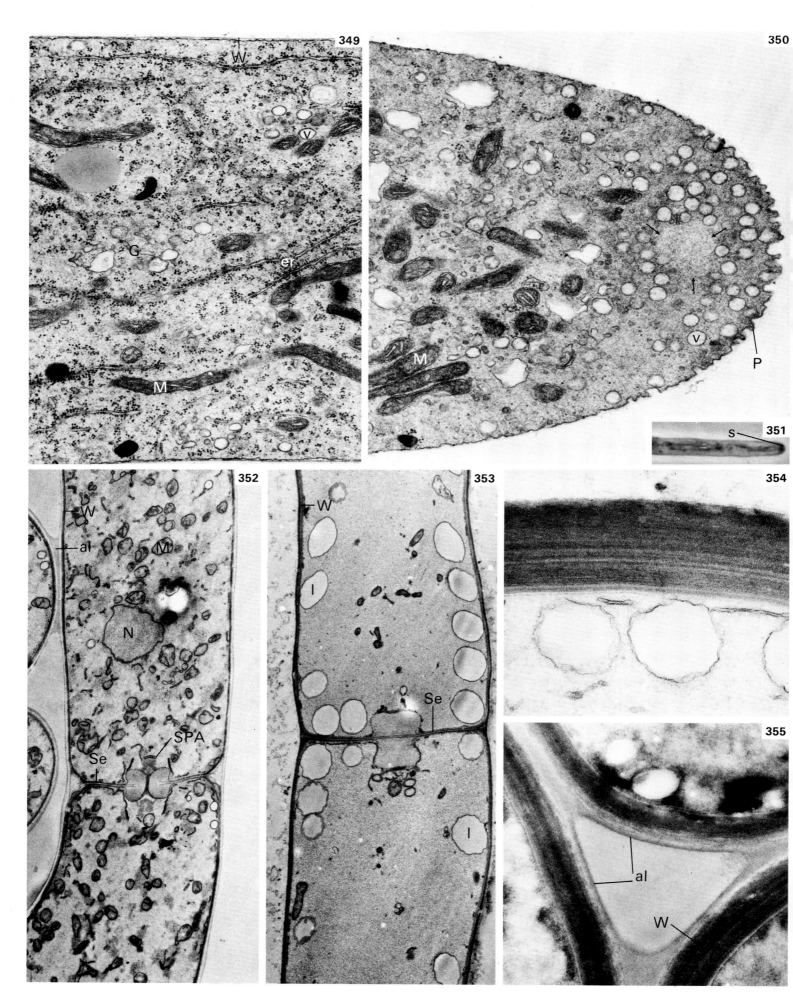

A. Vegetative Structures

Hyphae I. Cell wall synthesis and hyphal growth

The vegetative phase of Basidiomycotina is typically a hypha, although budding yeast-like cells are also formed in some genera. Apical growth of hyphae is believed to result from the fusion of **'wall vesicles'** apparently derived from the Golgi apparatus, with the apical wall (figs. 349, 350) (see also Section I, p. 5; Section II, p. 79). These vesicles are thought to contain wall compounds and enzymes involved in **wall synthesis**. In addition, the vesicles may contribute new **plasma membrane** to the growing hypha, and enzymes for maintaining **wall extensibility**, or they may be involved in **export** of enzymes from the cell. Wall vesicles are involved in the growth of clamp connections and have been implicated in bud formation in heterobasidiomycetous yeasts (fig. 386). They are also involved in basidiospore formation (fig. 402).

An **apical body ('Spitzenkörper')** is seen in growing hyphal apices, but disappears if growth is interrupted. By correlating light and electron microscope studies, the Spitzenkörper in *Armillaria mellea* (Vahl ex Fr.) Quélet has been demonstrated to be a spherical central region of the hyphal apex from which vesicles are absent (figs. 350, 351).

Hyphae show a progressive differentiation of cytoplasm and walls along their length. In *A. mellea* the apical region of the growing hypha contains numerous vesicles but few ribosomes, and other organelles are excluded from the immediate apex (fig. 350). In the sub-apical region, 30 μm behind the apex, are found numerous ribosomes, rough endoplasmic reticulum, Golgi cisternae and associated vesicles, along with mitochondria (fig. 349). Mature hyphae of *Rhizoctonia solani* Kuehn contain dense protoplasm and have thin cell walls (fig. 352). Old hyphae contain few cytoplasmic organelles, may contain lipid droplets and have thickened, multilayered walls (figs. 353, 354). Sclerotia have thickened walls with an amorphous material on the outer surface similar to that found on mature hyphae (compare figs. 352, 355).

Additional reading

BUTLER, E. E. and BRACKER, C. E. (1970). Morphology and cytology of *Rhizoctonia solani*, pp. 32–51. In: *Rhizoctonia solani*: Biology and Pathology (J. R. Parmenter, Jr, ed), University of California Press, Berkeley.

GROVE, S. N. and BRACKER, C. E. (1970). Protoplasmic organization of hyphal tips among fungi: Vesicles and Spitzenkörper. *J. Bact.*, **104**, 989.

Fig. 349
Part of the sub-apical zone of a hypha of *Armillaria mellea* (Vahl ex Fr.) Quélet showing clusters of vesicles (v) similar to those found at the hyphal apex. The clusters of vesicles are associated with Golgi cisternae or tubules (G) in zones of cytoplasm which contain few ribosomes compared to the surrounding protoplasm. Mitochondria (M) and endoplasmic reticulum (er) occur throughout this region. Compare the concentration of ribosomes with that in the apical region. (W) thin lateral wall of hypha. Glutaraldehyde—osmium tetroxide fixation. × 29,000.

Fig. 350
The apical zone of a hypha showing the vesicles (v) clustered at the hyphal tip, and the central region that is free from vesicles (arrows) which corresponds to the Spitzenkörper seen with the light microscope. The irregular profile of the plasma membrane (P) and the wall in the hyphal apex suggests vesicular fusion with the plasma membrane. Glutaraldehyde—osmium tetroxide fixation. × 34,000.

Fig. 351
The apex of a living hypha of *A. mellea* observed with phase-contrast microscopy showing the Spitzenkörper in the hyphal tip. × 2,500.

Fig. 352
Longitudinal section through a region of a mature hypha (about 24 hours old) of *Rhizoctonia solani* Kuehn. One of the several nuclei (N) in each cell and numerous mitochondria (M) are evident. The septal pore apparatus (SPA) connects adjacent cells. An amorphous layer (al) is present on the surface of the hyphal walls (W). (Se) septum. × 6,500.

Fig. 353
Longitudinal section through an aged hypha with few cytoplasmic organelles and numerous lipid droplets (l). The section is non-median and does not show the septal pore. (W) lateral wall; (Se) septum. × 8,200.

Fig. 354
Part of an older, brown hypha showing the multilayered wall. × 36,000.

Fig. 355
The amorphous layer (al) between the multilayered walls (W) of several cells of the sclerotium. × 34,000.

Figs. 349–55
(*352–5 Potassium permanganate fixation.*) *349, 350 From* GROVE, S. N. and BRACKER, C. E. (1970). *J. Bact.*, **104**, 989. *351 Micrograph by* DRS S. N. GROVE and C. E. BRACKER, Purdue University, Lafayette, Indiana. *352–5 From* BUTLER, E. E. and BRACKER, C. E. (1970) in *Rhizoctonia solani: Biology and Pathology* (J. R. Parmenter, Jr., ed.), originally published by the University of California Press, Berkeley. Reprinted by permission of the Regents of the University of California.

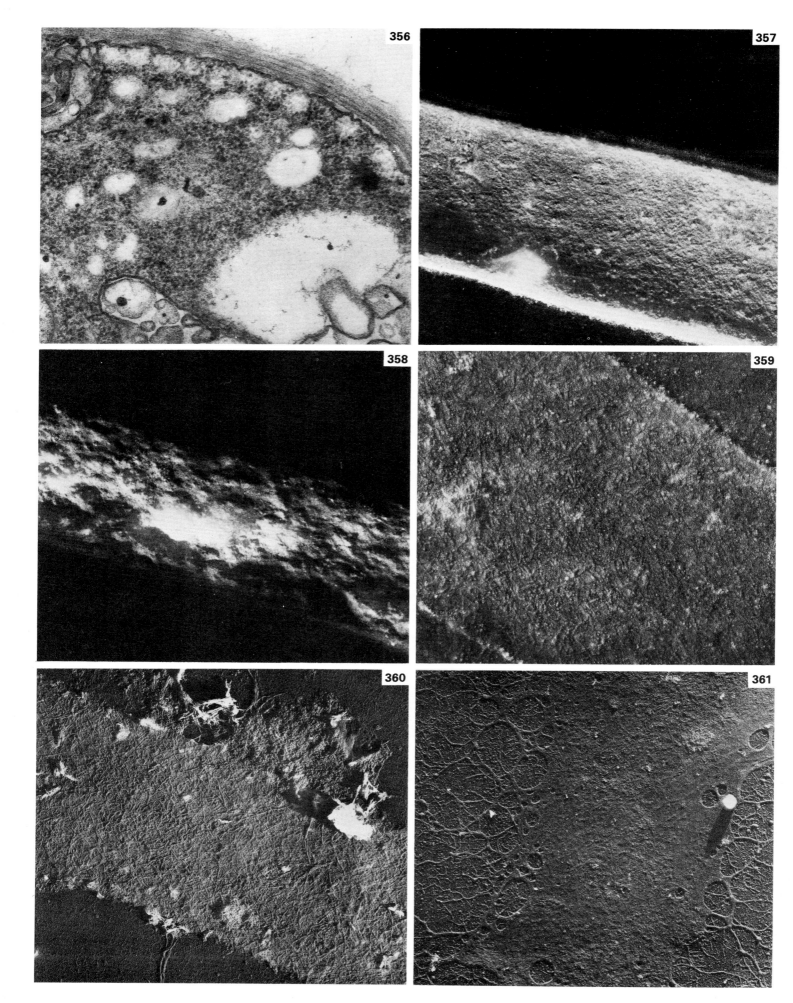

A. Vegetable Structures (cont.)

Hyphae II. Cell wall structure

The enzymic-dissection method employed with *Schizophyllum commune* Fr. is similar to that described for *Phytophthora parasitica* (Section I, p. 9) and *Neurospora crassa* (Section II, p. 81). *Schizophyllum* walls show a four-layered structure in section (fig. 356) and are known to contain $\beta 1,3$; $\beta 1,6$ linked **glucan**, an $\alpha 1,3$ linked glucan, **protein** and **chitin**. The enzymes employed are **laminarinase** ($\beta 1,3$; $\beta 1,6$ glucanase), **pronase** and **chitinase**. Control hyphae incubated in buffer show only an amorphous, roughened appearance in shadow cast preparations (fig. 357). This appearance does not alter after single separate treatments with laminarinase, pronase or chitinase. However when treated with a 5% KOH solution, the outer wall layer is removed. This layer can be shown by infrared spectroscopy to contain α-linked glucan. After the KOH treatment, hyphal surfaces appear irregularly roughened (fig. 358). Potassium hydroxide-treated material, after thorough washing in buffer, may be further subjected to enzymic degradation. For example, if KOH treatment is followed by laminarinase, the outer roughened amorphous region is removed, revealing a finely granular amorphous layer, under which the outlines of microfibrils can be distinguished (fig. 359). If the KOH/laminarinase treatment is followed by pronase then this inner amorphous layer is removed revealing the randomly oriented chitin microfibrils (fig. 360). The sequential treatment KOH/laminarinase/pronase/chitinase results in virtual dissolution of the microfibrils (fig. 361).

Additional reading

BACON, J. S. D., JONES, D., FARMER, V. C. and WEBLEY, D. M. (1968). The occurrence of $\alpha 1,3$ glucan in *Cryptococcus, Schizosaccharomyces* and *Polyporus* species and its hydrolysis by a *Streptomyces* culture filtrate lysing cell walls of *Cryptococcus. Biochim. Biophys. Acta*, **158**, 313.
HUNSLEY, D. and BURNETT, J. H. (1970). The ultrastructural architecture of the walls of some hyphal fungi. *J. gen. Microbiol.*, **62**, 203.
WESSELS, J. G. H. (1965). Morphogenesis and biochemical processes in *Schizophyllum commune. Wentia*, **13**, 1.

Fig. 356

Section through part of a hypha of *Schizophyllum commune* Fr. showing the four layers of the wall. Glutaraldehyde–osmium tetroxide fixation. × 45,000.

Fig. 357

Shadowed preparation of cell walls of *S. commune* after incubation in buffer only. Pd/Au shadowed 40/60 Cot$^{1-3}$. × 20,000.

Fig. 358

Cell wall preparation after treatment with 5% KOH solution. Pd/Au shadowed 40/60 Cot$^{1-3}$. × 20,000.

Fig. 359

Cell wall preparation after treatment with KOH followed by laminarinase. Note microfibril outlines beneath amorphous layer. Pd/Au shadowed 40/60 Cot$^{1-3}$. × 22,500.

Fig. 360

Cell wall preparation after KOH/laminarinase/pronase treatment. Random microfibrils can now be seen. Pd/Au shadowed 40/60 Cot$^{1-3}$. × 20,000.

Fig. 361

Cell wall preparation after KOH/laminarinase/pronase/chitinase treatment. Microfibrils have been almost completely removed. Pd/Au shadowed 40/60 Cot$^{1-3}$. × 15,000.

Figs. 356–61 *356, 358, 359, 361 From* HUNSLEY, D. and BURNETT, J. H. (1970). *J. gen. Microbiol.*, **62**, 203. *357, 360 Micrographs by* DR D. HUNSLEY, Department of Agricultural Science, University of Oxford.

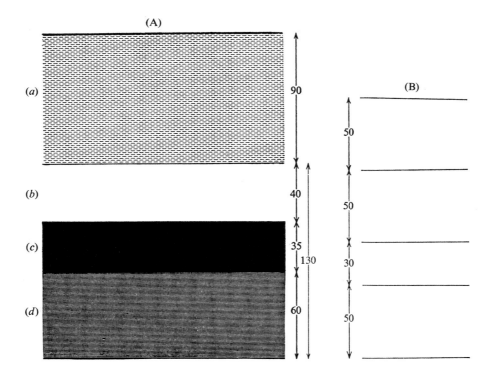

A. Vegetative Structures (cont.)

Hyphae II. Cell wall structure (cont.)

Fig. 362

(A). Diagrammatic reconstruction of section through the wall of a hypha from a 5-day culture of *S. commune*, based on KOH- and enzymic-dissection experiments. The numbers represent the mean thickness of the layers in nm. (*a*) outermost, amorphous glucan layer, possibly containing α1,3 linkages and soluble in cold KOH; (*b*) amorphous glucan containing β1,3 and β1,6 linkages; (*c*) a discrete layer of protein; (*d*) innermost layer of

chitin microfibrils possibly intermixed with protein.
(B). Layers visible in a section of an untreated wall from a 5-day culture fixed in glutaraldehyde–osmium tetroxide. Numbers represent the mean thickness of the layers in nm.

Fig. 362 *From* HUNSLEY, D. and BURNETT, J. H. (1970). *J. gen. Microbiol.*, **62**, 203.

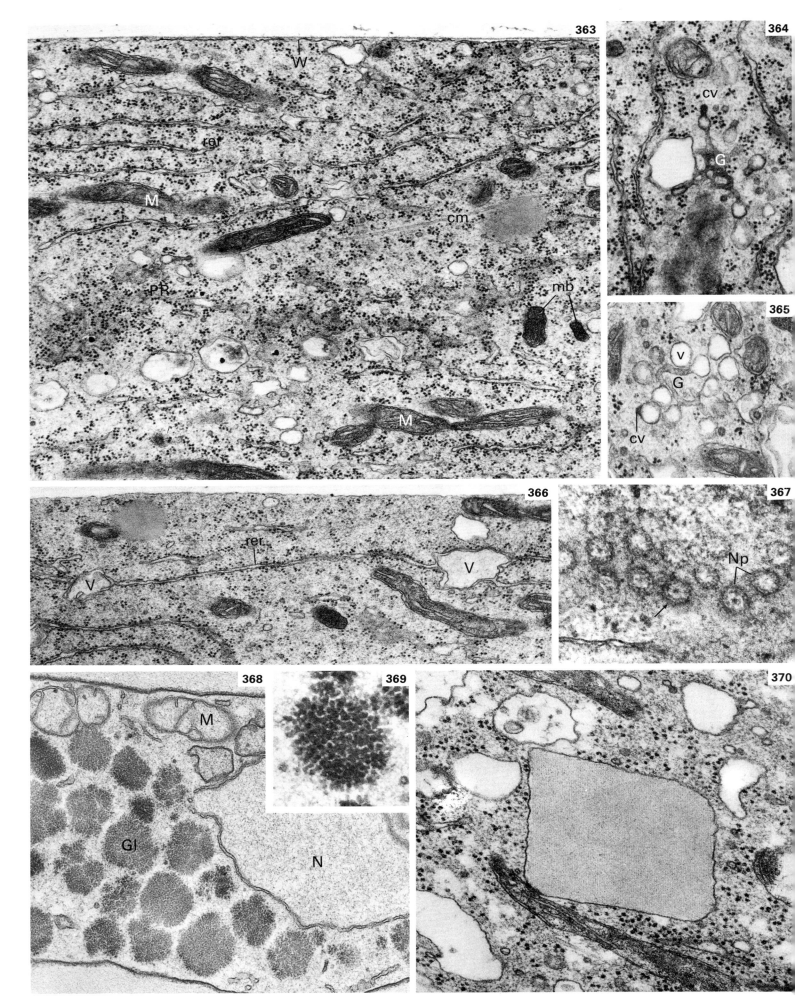

A. Vegetative Structures (cont.)

Hyphae III. Cell components

The cytoplasm of Basidiomycotina contains the usual complement of cellular organelles found in other eukaryotic cells. However, the **Golgi apparatus**, like that in some other groups of fungi, is distinctive as is the development of the endoplasmic reticulum around the septum.

The membrane system within the cell includes membranes of the nuclear envelope, endoplasmic reticulum, Golgi apparatus and secretory vesicles, and the plasma membrane. These cell components are structurally or functionally interrelated. The **nuclear envelope** possesses pores or annuli which are eight-sided and contain a central granule (fig. 367). The **endoplasmic reticulum** is sometimes attached to the nuclear envelope. In the cytoplasm, endoplasmic reticulum is frequently seen as flattened sacs or cisternae which may bear ribosomes on the external surface (fig. 363). A surface view of such a cisternum shows the ribosomes aggregated in polyribosomal arrays on which protein synthesis occurs. Numerous ribosomes remain free in the cytoplasm. Endoplasmic reticulum is common in young, but not in older hyphae. Coated vesicles are thought to form on the endoplasmic reticulum and to be involved in the formation of **Golgi cisternae** (fig. 364). The Golgi apparatus in Basidiomycotina is composed of individual cisternae which are not organized into cisternal stacks or dictyosomes (figs. 364, 365) (see also Section II, p. 85). Vesicles which arise from the Golgi cisternae function in the export of material from the cell. The vesicles migrate to and fuse with

the hyphal apex or growing region, and the vesicle membranes contribute to the formation of new plasma membrane (fig. 350).

Rough endoplasmic reticulum is associated with forming **vacuoles** in young hyphae of *Armillaria mellea* (Vahl ex Fr.) Quélet, and may be involved in vacuolar initiation (fig. 366). Vacuoles have been reported to contain hydrolytic enzymes in basidiocarps of *Coprinus* and to correspond functionally to **lysosomes**. **Microbodies** are organelles which range in size from about 0·2 to 1·5 μm in diameter and have granular contents (figs. 363, 403). Their membrane is similar in thickness to that of the endoplasmic reticulum to which they may be joined. They sometimes contain crystalline inclusions. Their precise role in Basidiomycotina is not known. Membrane-bound crystalline inclusions in the cytoplasm (fig. 370) may be related to microbodies. Proteinaceous crystals have been reported within nuclei.

Other cytoplasmic components include **microtubules** which are about 20–25 nm in diameter and are similar in size and structure to those in the nuclear spindle (figs. 363, 401, 403). **Mitochondria** typically contain plate-like cristae (figs. 363, 386). **Glycogen** is a common storage product and is found in non-apical regions of the hyphae and in reproductive cells. It appears as aggregates of particles 15 to 30 nm in diameter (figs. 368, 369). **Lipid** droplets are common in mature cells or spores (figs. 353, 399, 400, 412) but are usually absent from growing regions of the hyphae.

Additional reading

BRACKER, C. E. (1967). Ultrastructure of fungi. *A. Rev. Phytopathol.*, **5**, 343.
MOORÉ, D. J., MOLLENHAUER, H. H. and BRACKER, C. E. (1971). Origin and continuity of Golgi apparatus. In: *Results and Problems in Cell Differentiation* II. *Origin and Continuity of Cell Organelles* (J. Reinert and H. Ursprung, eds.), pp. 82–126. Springer-Verlag, Berlin.

Fig. 363

Young hypha of *Armillaria mellea* (Vahl ex Fr.) Quélet containing extensive rough endoplasmic reticulum (rer). Numerous polyribosomes (PR) are seen in surface views of the cisternae of the endoplasmic reticulum. Microbodies (mb), mitochondria (M) and microtubules (cm) are also evident. (W) lateral wall of hypha. Glutaraldehyde–osmium tetroxide fixation. × 37,000.

Figs. 364, 365

Smooth-surfaced, fenestrated or tubular Golgi cisternae (G) in the sub-apical region of hyphae of *A. mellea*. Each cisterna lies in a zone of cytoplasm which contains few ribosomes compared to the surrounding cytoplasm. Vesicles (v) similar to those in the hyphal apex, and coated vesicles and protuberances (cv) are associated with the Golgi cisternae. Glutaraldehyde–osmium tetroxide fixation. × 55,000.

Fig. 366

Hypha of *A. mellea* showing the relationship between the endoplasmic reticulum and vacuole initiation. Rough endoplasmic reticulum (rer) is continuous with double-membrane-bounded cavities that seem to be vacuolar precursors (V). Glutaraldehyde–osmium tetroxide fixation. × 37,000.

Fig. 367

Nuclear pores or annuli in a tangential section of the nuclear

envelope from a basidium of *Boletus rubinellus* Peck. The nuclear pore contains a central granule and the perimeter of the pore is angular (arrow). Glutaraldehyde–osmium tetroxide fixation. × 94,000.

Fig. 368

Subhymenial cell of *B. rubinellus* with numerous clusters of glycogen particles (Gl). (M) mitochondrion, (N) nucleus. Potassium permanganate fixation. × 26,000.

Fig. 369

High magnification of glycogen in a subhymenial cell of *B. rubinellus*. Glycogen is present in the form of approximately isodiametric beta particles which are aggregated into large clusters termed alpha particles. Potassium permanganate fixation. × 74,000.

Fig. 370

A crystalline body, common in the hyphae of *A. mellea*. Glutaraldehyde–osmium tetroxide fixation. × 37,000.

Figs. 363–70 *363, 366, 370 Micrographs by* DRS S. N. GROVE and C. E. BRACKER, Purdue University, Lafayette, Indiana. *364, 365 From* GROVE, S. N. and BRACKER, C. E. (1970). *J. Bact.*, **104**, 989. *367–9 Micrographs by* D. J. MCLAUGHLIN, University of Minnesota, St Paul.

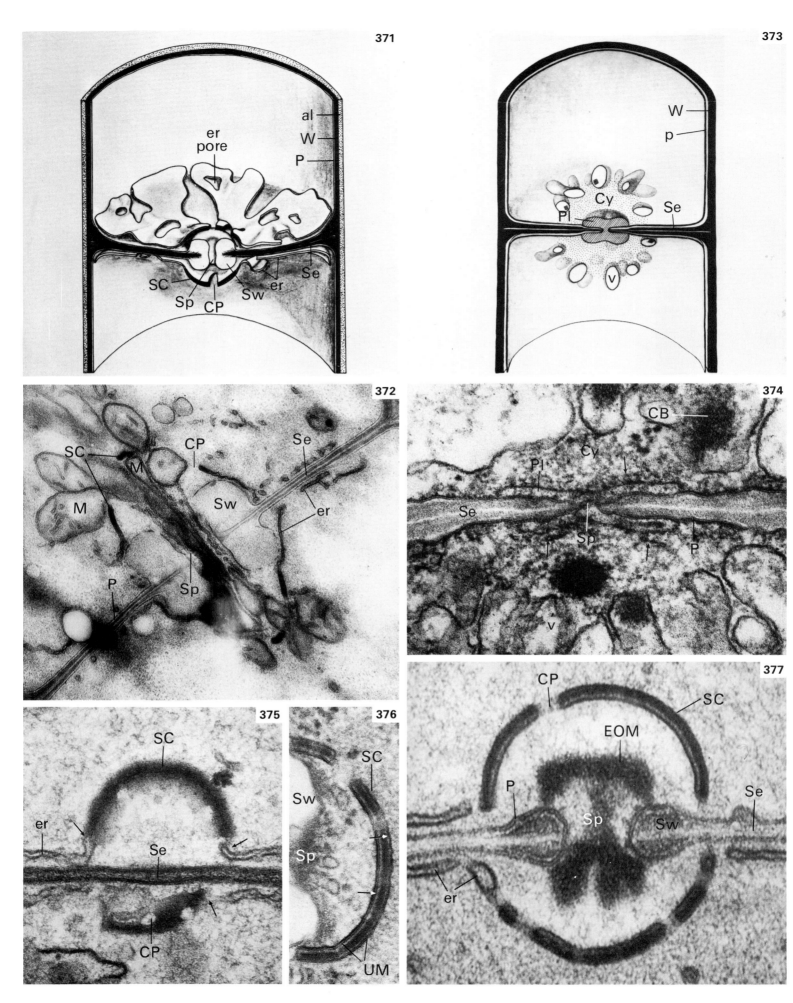

A. Vegetative Structures (cont.)

Septa and associated structures

The septal pore region appears to be more elaborate in certain Basidiomycotina than in any other Fungi. There are two basic types of septa encountered in the Basidiomycotina: that of the rusts and other Teliomycetes, and that of the Hymenomycetes and Gasteromycetes. In the rusts the septa are either perforate with an open central pore, or the septal pore is blocked by a plug which seems to prevent cytoplasmic continuity between adjacent cells (figs. 373, 374). A zone of cytoplasm, typically free of organelles, surrounds the septal pore. This rust-type septum resembles that of the Ascomycotina. The septal pore with its associated structures is collectively known as the **septal pore apparatus**. In the Hymenomycetes (here used to include Phragmobasidiomycetidae and Holobasidiomycetidae) and Gasteromycetes, the septal pore apparatus consists of the **septal pore swelling** (**dolipore**) surrounding the pore which provides cytoplasmic continuity between adjacent cells (figs. 371, 372, 377). The plasma membrane of adjacent cells is continuous through the pore (fig. 377). **Septal pore caps** (**parenthesomes**) enclose the swellings which project into both cells.

The septum arises by the centripetal growth of a cross wall from the lateral wall of the hypha. The septal pore swellings are formed on the perimeter of the pore from material which seems to be of different composition from that of the cross wall. The septal pore cap is continuous with and probably derived from endoplasmic reticulum. However, it is a highly modified form of **annulate endoplasmic reticulum** and appears to have a rigid structure (figs. 375, 377). Three types of septal pore cap are known: (1) a solid cap without pores; (2) a cap with large apertures (up to 1 μm diam.) as in *Rhizoctonia solani* Kühn (figs. 371, 372); and (3) a cap with small pores (40–80 nm diam.) as in *Boletus rubinellus* Peck (figs. 375–377). The latter type has also been reported in some Gasteromycetes. The type of cap found in *R. solani* does not prevent the intercellular movement of large organelles, but the other types of caps may do so. However, under certain circumstances nuclei may move between cells with small-pored caps. During dikaryotization in *Coprinus lagopus* Fr. nuclei pass from cell to cell after the breakdown of the septal pore apparatus.

A plugging precursor which may serve to block the pore on the death or disruption of the cell may be found within the septal pore (fig. 377).

Additional reading

BRACKER, C. E. (1967). Ultrastructure of fungi. *A. Rev. Phytopathol.*, **5**, 343.
BUTLER, E. E. and BRACKER, C. E. (1970). Morphology and cytology of *Rhizoctonia solani*. In: *Rhizoctonia solani*: Biology and Pathology. (J. R. Parmenter, Jr. ed.), pp. 32--51. University of California Press, Berkeley.
LITTLEFIELD, L. J. and BRACKER, C. E. (1971). Ultrastructure of septa in *Melampsora lini. Trans. Br. mycol. Soc.*, **56**, 181.

Figs. 371–377

Abbreviations: (al) amorphous layer; (Cy) cytoplasm around septal pore, usually organelle-free; (CB) crystalline body; (CP) pore in septal pore cap; (Se) septum; (W) lateral wall; (M) mitochondrion; (EOM) electron-opaque material; (er) endoplasmic reticulum; (Pl) septal plug; (P) plasma membrane; (Sw) septal swelling; (SC) septal pore cap; (Sp) septal pore; (UM) unit membrane; (v) vesicle.

Fig. 371

A drawing of the septal pore apparatus of *Rhizoctonia solani* Kühn. The open pore (Sp) surrounded by a doughnut-shaped swelling (Sw) provides cytoplasmic continuity between adjacent cells. A septal pore cap (SC) with large discontinuities (CP) surrounds the pore swellings and is continuous with the endoplasmic reticulum (er) which borders the septum (Se).

Fig. 372

Longitudinal section through the septal pore apparatus of *R. solani*. Notice the large size of the pores (CP) in the septal pore caps (SC) and the mitochondria (M) streaming through the pore (Sp). The endoplasmic reticulum (er) is continuous with the pore caps (SC). Potassium permanganate fixation. × 27,000.

Fig. 373

A drawing of the septum in the uredial thallus of the rust *Melampsora lini* (Ehrenb.) Lév. The pore in the septum is blocked by a pulley wheel-shaped plug (Pl) which is usually surrounded by a matrix of dense cytoplasm (Cy) from which organelles are excluded. This cytoplasmic matrix is usually surrounded by vesicles (v) or microbodies containing crystalline inclusions.

Fig. 374

A perforate septum in a mature hypha of *M. lini*. The electron-dense cytoplasm (Cy) bordered by vesicles (v) is present on both sides of the septum (Se). The electron-opaque band forming the septal plug (Pl) is complete over the pore (Sp). Near its periphery, in places, is a faint, electron-opaque secondary band (arrows). Glutaraldehyde–osmium tetroxide fixation. × 108,000.

Fig. 375

A section of a sub-hymenial septum of *Boletus rubinellus* Peck showing a tangential view of the septal pore cap (SC) with small, round pores (CP). The cap is continuous with the endoplasmic reticulum (arrows). Potassium permanganate fixation. × 74,000.

Fig. 376

The septal pore cap (SC) consists of a unit membrane (UM) surrounding an electron-opaque layer (upper arrow) and electron-transparent layers. On its outer surface toward the septal pore is a narrow electron-opaque layer (lower arrow). Glutaraldehyde–osmium tetroxide fixation. × 110,000.

Fig. 377

The septal pore apparatus, probably from a young subhymenial cell. The small septal pore swellings (Sw) contain electron-opaque material adjacent to the cross wall (Se). The plasma membrane (P) is continuous between adjacent cells. Continuity of the endoplasmic reticulum (er) and cap is evident. The septal pore (Sp) contains electron-opaque material (EOM) connected to the swellings (Sw) which may represent a plugging precursor. Potassium permanganate fixation. × 105,000.

Figs. 371–7 *371 From* BRACKER, C. E. and BUTLER, E. E. (1963). *Mycologia*, **55**, 35. *372 From* BUTLER, E. E. and BRACKER, C. E. (1970), in *Rhizoctonia solani: Biology and Pathology* (J. R. Parmenter, Jr., ed.), University of California Press, Berkeley; reprinted by permission of The Regents of the University of California. *373, 374 From* LITTLEFIELD, L. J. and BRACKER, C. E. (1971). *Trans. Br. mycol. Soc.*, **56**, 181. *375–7 Micrographs by* D. J. MCLAUGHLIN, University of Minnesota, St Paul.

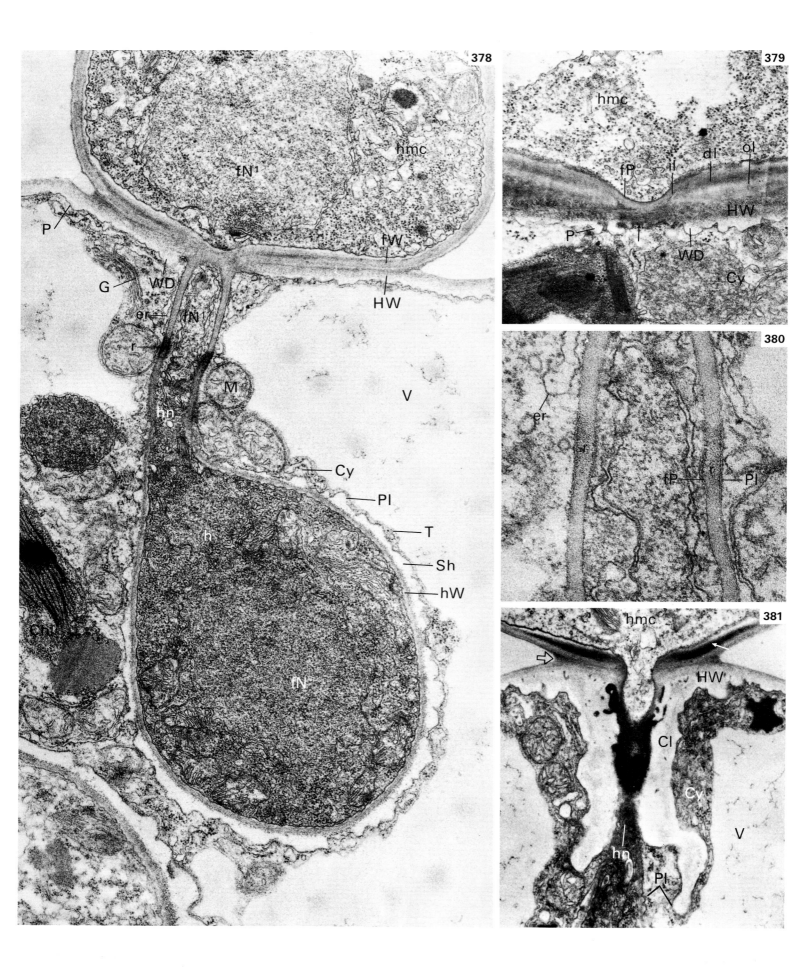

A. Vegetative Structures (cont.)

Haustoria

Rusts and some other Basidiomycotina form haustoria. The protoplast of the fungal parasite is not in direct contact with the host cytoplasm (see also Section I, p. 21; Section II, p. 91). In the flax parasite *Melampsora lini* (Ehrenb.) Lév. the cytoplasm of the haustorium is separated from the host cytoplasm by the fungal wall, a **sheath** (**encapsulation**), and the host plasma membrane and usually contains one or two nuclei (fig. 378). The haustorium arises from a **mother cell** which is in contact with the host wall. A layer of material forms between the wall of the host and mother cell and may aid in adhesion (fig. 381). During haustorial development and probably before haustorial initiation a new wall layer forms within the mother cell wall where it contacts the host wall. The **penetration peg** which enters the host cell develops from the inner wall layer or layers of the mother cell (fig. 379).

The plasma membrane of the host cell is invaginated by the developing haustorium and the host cell forms wall deposits around the site of penetration (figs. 378, 379). The host cell may form extensive **wall deposits** which can result in the formation of a **collar** around the neck of the haustorium (fig. 381). Usually the host wall or collar does not completely surround the haustorium. Studies in other rusts confirm that the host plasma membrane is invaginated by the invading haustorium and forms a continuous membrane around it.

The wall of the **haustorial neck** undergoes a series of changes from the site of penetration to the region of the sheath (figs. 378, 380). It consists of three distinctive zones: (1) the zone which arises from and resembles the mother cell wall; (2) a ring of dark-staining wall material below the first zone; and (3) the lower haustorial neck. The latter zone is surrounded by the sheath and is similar to the wall of the haustorium proper. The sheath does not surround the upper neck or the ring of dark-staining wall material. This ring has been found in other rusts. The changes in the haustorial wall appear to be related to the interaction of the haustorium with the host in the region of the sheath.

The technique of freeze-etching has been successfully used in studies on rust haustoria. This technique is useful for several reasons. It provides confirmation of the results obtained with fixed and embedded tissues while avoiding most or all chemical treatments for specimen preparation. It reveals the morphology of cell and organelle surfaces. Membrane surfaces usually bear particles whose sizes are characteristic for particular cell components and can be used to demonstrate the interrelationships between components.

The results obtained with frozen-etched haustoria of *M. lini* confirm those obtained with fixed and embedded material (see fig. 378) and provide new information on the nature of the host plasma membrane where it borders the haustorial sheath. It is in the region of the sheath that exchange of material between host and parasite presumably occurs. The surface of the host plasma membrane is relatively smooth over the haustorial neck while it is highly folded over the haustorium (fig. 382). The undulating nature of the host plasma membrane is also evident in the cross fracture through the haustorium in which the

Fig. 378

A longitudinal non-median section through the haustorial mother cell (hmc) and haustorium (h) of *Melampsora lini* (Ehrenb.) Lév., at the point where it penetrates the cell wall (HW) of its host, flax, *Linum usitatissimum* L. The haustorium and haustorial neck (hn) are surrounded by the invaginated host plasma membrane (Pl). One nucleus (fN) is in the haustorium, and the other (fN¹) appears to be partly in the haustorial neck and partly in the haustorial mother cell. An electron-lucent sheath (Sh) separates the host cytoplasm (Cy) from the haustorial wall (hW) around the body of the haustorium. The sheath is absent from the haustorial neck. A dark-staining ring (r) of wall material occurs in the haustorial neck. (fW), fungal wall. Host structures: (Chl) chloroplast; (G) Golgi dicytosome; (er) endoplasmic reticulum; (M) mitochondrion; (P) plasma membrane; (T) tonoplast; (V) vacuole; (WD) deposit on cell wall around fungal penetration. × 27,000.

Fig. 379

An early stage in the penetration of a mesophyll cell by a haustorial mother cell (hmc). The inner layer (il) of the fungal wall is evaginated through the outer portions and contacts the host wall (HW). The outer portion of the fungal wall includes the electron-dense middle layer (dl) and the outer layer (ol). The host wall stains more intensely at the site of penetration (arrow) than in other regions. Wall deposits (WD) are on the inner surface of the host wall around the penetration site. (fP) fungal plasma membrane; (P) host plasma membrane; (Cy) host cytoplasm. × 33,000.

Fig. 380

A high magnification micrograph of the midregion of the haustorial neck (see fig. 378). The ring (r) of wall material stains darker and is slightly thicker than the wall along the rest of the neck. The staining of both host and fungal plasma membranes (Pl, fP) is enhanced along the surface of the neck ring. Note the host endoplasmic reticulum (er) associated with the haustorial neck. × 90,000.

Fig. 381

A collar (Cl) of wall material continuous with the host cell wall (HW) surrounds the haustorial neck (hn). Note the electron-opaque layer (arrow) of the wall of the haustorial mother cell (hmc) and the layer of material (open arrow) between host and fungus walls. (Cy) host cytoplasm; (Pl) host plasma membrane; (V) host vacuole. × 28,000.

Figs. 378–81 *Simultaneous glutaraldehyde/osmium tetroxide fixation. From* LITTLEFIELD, L. J. and BRACKER, C. E. (1972). *Protoplasma,* **74**, 271.

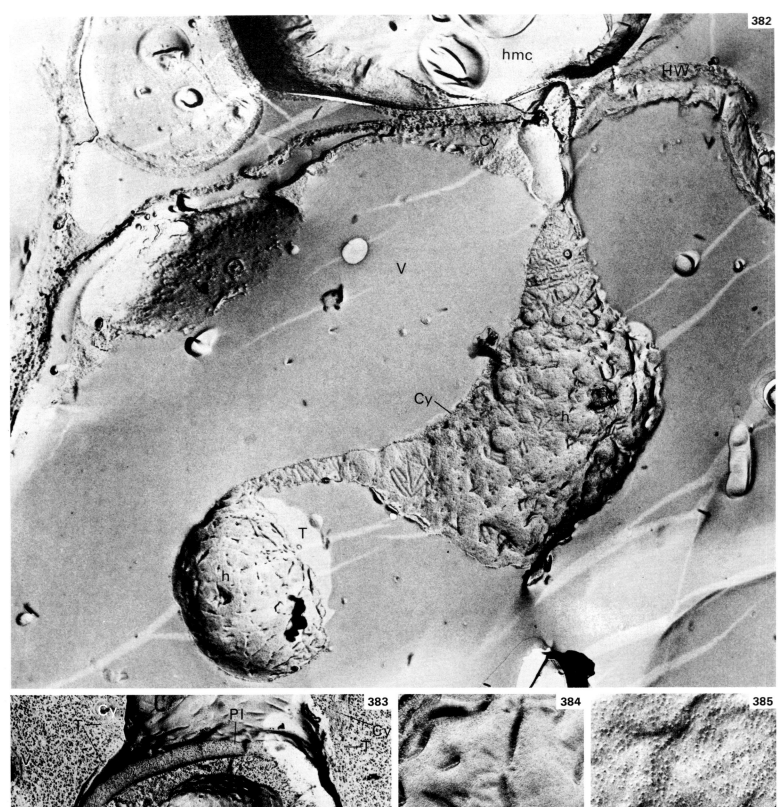

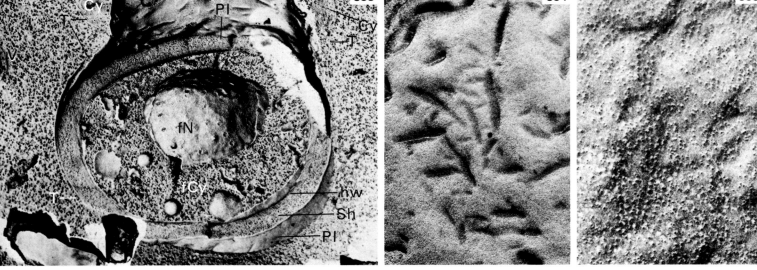

A. Vegetative Structures (cont.)

Haustoria (cont.)

haustorial cytoplasm and a small part of its non-folded plasma membrane can also be seen (fig. 383). The sheath appears coarse after freeze-etching while the haustorial wall is smoother. A thin band of host cytoplasm encloses the haustorium .Where the host plasma membrane borders

the haustorium it is folded to form elongate depressions and is smooth (fig. 384). Areas of the host plasma membrane other than those which surround the haustorium are not folded and bear granules on the surface (fig. 385).

Additional reading

COFFEY, M. D., PALEVITZ, B. A. and ALLEN, P. J., (1972). The fine structure of two rust fungi, *Puccinia helianthi* and *Melampsora lini. Can. J. Bot.*, **50**, 231.

LITTLEFIELD, L. J. and BRACKER, C. E. (1972). Ultrastructural specialization at the host–pathogen interface in rust-infected flax. *Protoplasma*, **74**, 271.

Fig. 382

The haustorium (h) arising from the haustorial mother cell (hmc) of *Melampsora lini* (Ehrenb.) Lév. is seen within a leaf cell of flax. The invaginated host plasma membrane covers the haustorium and is characterized by undulations and depressions. A thin layer of host cytoplasm (Cy) borders the lower, but not the upper, end of the haustorium. (V) host vacuole; (HW) host wall. Freeze-etched replica. × 16,000.

Fig. 383

Cross-fractured haustorium showing the fungal cytoplasm (fCy), and a nucleus (fN). The sheath (Sh) has a coarser texture after freeze-etching than the haustorial wall (hw). The tonoplast (T), which encloses a thin layer of host cytoplasm (Cy), can be

traced around most of the haustorium. Note the folds in the host plasma membrane (Pl). (V) host vacuole. Freeze-etched replica. × 18,000.

Fig. 384

Face view of host plasma membrane invaginated around a haustorium. Note the lack of granularity and the cleft-like depressions in the surface. Freeze-etched replica. × 70,000.

Fig. 385

Face view of non-invaginated host plasma membrane. Note the granules on the surface. Freeze-etched replica. × 70,000.

Figs. 382–5 *From* LITTLEFIELD, L. J. and BRACKER, C. E. (1972). *Protoplasma*, **74**, 271.

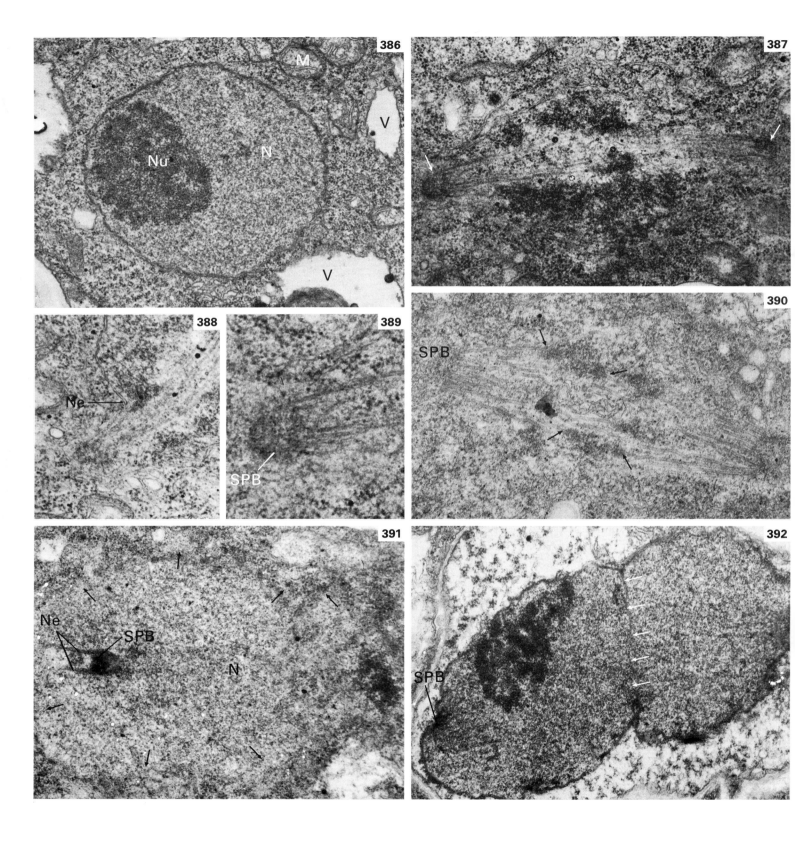

A. Vegetative Structures (cont.)

Mitosis

Nuclear division in Basidiomycotina is characterized by break down of the nuclear envelope during spindle formation. Two examples will be illustrated here.

In the rhizomorph of *Armillaria mellea* (Vahl ex Fr.) Kummer true mitosis occurs in which the nuclear envelope **breaks down** during prophase and chromosomes move apart with the assistance of a spindle apparatus. At interphase the **spindle pole body** is apparently absent (fig. 386). At prophase spindle pole bodies, which are about 400 nm in diameter, become visible outside the nuclear envelope and chromatin becomes evident within the nucleus. Later in division the spindle forms and its microtubules extend, through gaps in the nuclear envelope, towards the spindle pole bodies which are located in the cytoplasm (figs. 387–389). Finally the nuclear envelope breaks down completely and the chromosomes are connected to the spindle microtubules (fig. 380). After the chromosomes have reached the poles the nuclear envelope reforms, possibly under the direction of the spindle pole body (figs. 391, 392). The stages in mitosis are summarized in fig. 393.

Mitosis in *Polystictus versicolor* L. shows some differences from that in *A. mellea*. The nuclear envelope breaks down to allow the entrance of the spindle pole body into the nucleus. However, the enclosing membranes are then reorganized so that the nucleus and spindle pole bodies are surrounded by the nuclear envelope or cisternae of the endoplasmic reticulum during division. A spindle pole body is always present in *P. versicolor*, while it may be absent during interphase in *A. mellea*. Furthermore, in *P. versicolor* the spindle pole body is **bipolar** with a **middle piece** connecting the two **globular ends** during most of interphase (see figs. 395, 405). The two globular ends apparently separate to form individual spindle pole bodies as spindle microtubules develop between them. Each of the two

globular structures therefore occupies a pole of the mitotic spindle. After mitosis each spindle pole body forms a second globular end early in interphase to re-establish the bipolar structure.

The interruption of the nuclear envelope and the resulting direct contact of the spindle pole bodies with the nucleoplasm during mitosis is characteristic of Basidiomycotina and appears to be absent in most other Eumycota.

Basidiomycetous life cycles have recently been demonstrated for several **asporogenous yeasts**, and the type of mitosis in these yeasts has now been shown to be similar to that of mycelial Basidiomycotina. During bud formation at interphase in the heterobasidiomycetous yeast *Leucosporidium scottii* (*Candida scottii*) Diddens and Lodder, a **bipolar spindle pole body** is situated beside the nucleus on the side adjacent to the developing bud (figs. 394, 395). The nucleus then appears to grow into the bud, the nucleolus staying in the part of the nucleus which remains in the mother cell (fig. 396). One side of the nuclear envelope breaks down in the bud, and, when the spindle pole body enters the nucleus, its two **globular components** apparently separate to become the poles of the spindle and the spindle microtubules develop between them (fig. 397). The nucleolus disintegrates in the mother cell. As the spindle elongates one pole re-enters the mother cell, and daughter nuclei are reconstituted in bud and mother cell. Mitosis is carried out essentially within the bud with the nuclear envelope interrupted along one side. The stages of mitosis are summarized in fig. 398.

The type of mitosis found in *L. scottii* has also been found in two other heterobasidiomycetous yeasts and differs from that found in ascomycetous yeasts. Mitosis in the latter is intranuclear involving no break down in the nuclear envelope and division occurs in the neck region between mother and daughter cells (see Section II, p. 93).

Fig. 386

Spherical interphase nucleus (N) of *Armillaria mellea* (Vahl ex Fr.) Kummer with a nucleolus (Nu). (M) mitochondrion; (V) vacuole. × 25,000.

Fig. 387

Median section through a dividing nucleus. The spindle apparatus appears to originate from the spindle pole bodies (arrows) located at opposite ends of the nucleus. × 31,000.

Fig. 388

Nuclear envelope (Ne) which is broken where the spindle passes through it. × 36,000.

Fig. 389

The spindle microtubules are connected directly to the spindle pole body (SPB). × 51,000.

Fig. 390

Late metaphase/early anaphase of mitosis. Arrows indicate attachments of chromosomes to the spindle microtubules. Note the absence of a nuclear envelope. (SPB) spindle pole body. × 31,000.

Fig. 391

Post-telophase resynthesis of the nuclear envelope (Ne) appears to begin in the vicinity of the spindle pole body (SPB). Arrows indicate limits of nucleoplasm (N). × 32,000.

Fig. 392

Restoration of the nuclear envelope. Sister nuclei appear to separate by the formation of an intranuclear membrane (arrows). The spindle pole body (SPB) has become altered into a plaque-like structure. × 30,000.

Figs. 386–92 *Glutaraldehyde–osmium tetroxide fixation. From* MOTTA, J. J. (1969). *Mycologia*, **61**, 873.

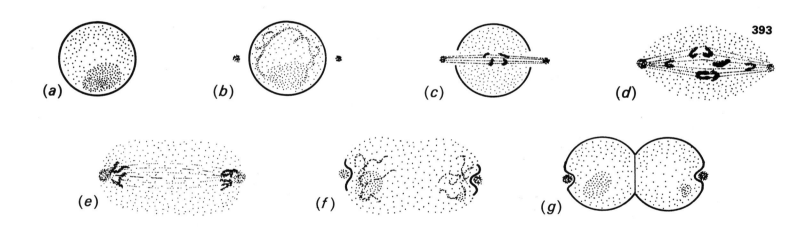

(a)

(b)

(c)

(d)

(e)

(f)

(g)

A. Vegetative Structures (cont.)

Mitosis (cont.)

Mitosis within the bud appears to be a feature of those yeasts which are related to Basidiomycotina. Before nuclear division the bipolar spindle pole body of *L. scottii*, like that of Hymenomycetes (fig. 405), consists of two globular ends connected by a **middle piece** and surrounded

by a **ribosome-free zone** (fig. 395). However, in some other heterobasidiomycetous yeasts and rusts the spindle pole body consists of two bar-shaped, rather than globular, components connected by a middle piece and resembles the spindle pole bodies of some Ascomycotina.

Additional reading

GIRBARDT, M. (1968). Ultrastructure and dynamics of the moving nucleus. *Symp. Soc. exp. Biol.*, **22**, 249.

MCCULLY, E. K. and ROBINOW, C. F. (1972). Mitosis in heterobasidiomycetous yeasts. I. *Leucosporidium scottii (Candida scottii). J. Cell Sci.*, **10**, 857.

MCCULLY, E. K. and ROBINOW, C. F. (1972). Mitosis in heterobasidiomycetous yeasts. II. *Rhodosporidium* sp. (*Rhodotorula glutinis*) and *Aessosporon salmonicolor* (*Sporobolomyces salmonicolor*). *J. Cell Sci.*, **11**, 1.

MOTTA, J. J. (1969). Somatic nuclear division in *Armillaria mellea. Mycologia*, **61**, 873.

Fig. 393(a)–(g)

Diagrammatic interpretation of somatic nuclear division in *A. mellea*. (*a*) interphase, (*b*) prophase: chromatin is peripherally distributed in the nucleus and the spindle pole bodies appear outside it. (*c*) metaphase: the spindle microtubules enter the nucleus through perforations in the nuclear envelope and the nucleolus begins to disappear. (*d*) anaphase: chromosomes are attached to the spindle microtubules and the nuclear envelope breaks down. (*e*), (*f*), (*g*) telophase: chromosomes disperse and the nucleoli reappear. Reformation of the nuclear envelope appears to begin at the spindle pole bodies.

Fig. 393 *Modified from* MOTTA, J. J. (1969). *Mycologia*, **61**, 873.

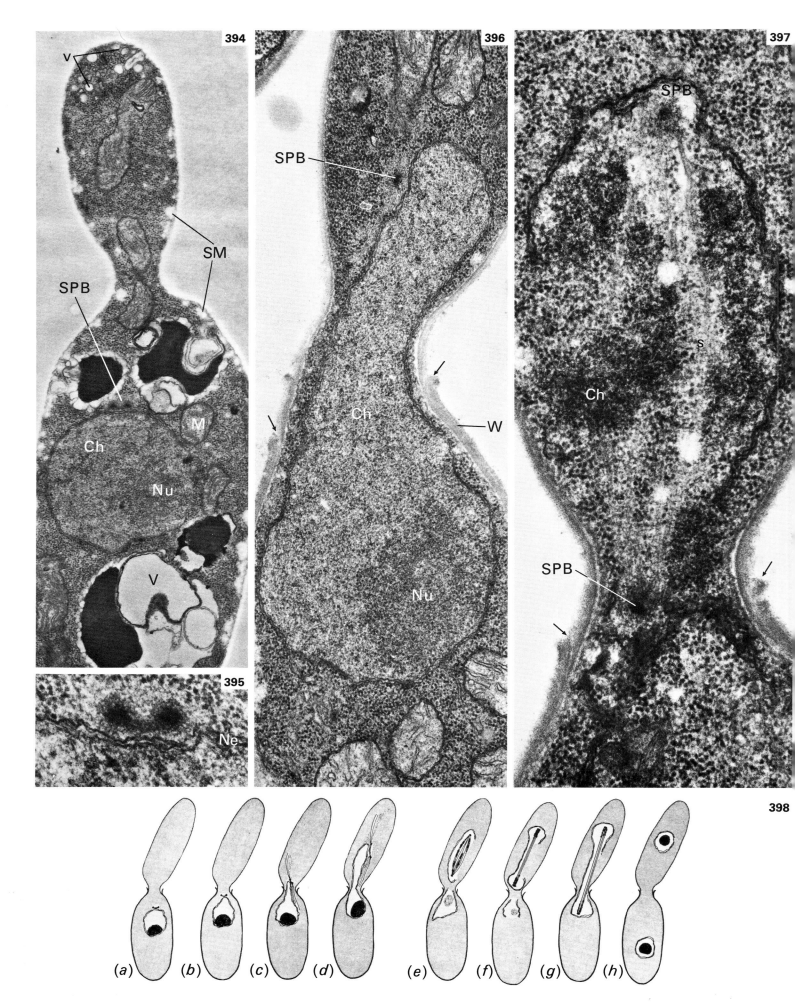

A. Vegetative Structures

Mitosis (cont.)

Fig. 394

A mother cell and young bud of the heterobasidiomycetous yeast *Leucosporidium scottii*. The nucleus, located toward the middle of the mother cell, is spherical and has a chromatin-containing region (Ch) of low electron-density and an electron-opaque nucleolus (Nu). In the cytoplasm are mitochondria (M), vacuoles (V), electron-transparent regions under the plasma membrane which seem to be the sites of storage material (SM), and a region of vesicles (v), near the tip of the bud. Outside the nuclear envelope is a bipolar spindle pole body (SPB). × 21,800.

Fig. 395

A higher magnification of a spindle pole body to show its two globular, electron-opaque ends and the bridge-like middle piece. The nucleus is partially visible at the bottom of the micrograph. Notice that the organelle is located close to but apparently not attached to the nuclear envelope (Ne). × 70,000.

Fig. 396

Movement of part of the nucleus into the bud before mitosis. Part of what is presumed to be chromatin (Ch) protrudes into the bud while the nucleolus (Nu) remains in the rounded base of the nucleus in the mother cell. One of the two electron-opaque spherical components of the spindle pole body (SPB) is visible near the tip of the nucleus. Notice that the cell wall (W) of the bud is thinner than that of the mother cell. Arrows at the base of the bud point to the outer layers of mother cell wall which have broken open during bud formation. × 36,800.

Fig. 397

A mitotic spindle (s) with a spindle pole body (SPB) visible at each pole. The nuclear envelope is present along the right side of the spindle but the left side is open to the cytoplasm. Masses of amorphous electron-opaque material interpreted as chromosomes (Ch) are scattered along the spindle. Arrows point to regions where the outer wall layers of the mother cell have broken open during bud formation. This feature shows that the spindle is located inside the bud. × 67,000.

Fig. 398 (a)–(h)

Diagrammatic interpretation of mitotic events in *L. scottii*. (a) interphase; (b), (c), (d) stages in the movement of a large portion of the nucleus into the bud. It is accompanied by a bipolar spindle pole body and cytoplasmic microtubules. (e), (f) spindle formation and chromosome separation inside the bud. Note that the nuclear envelope is not present along one side of the spindle. A portion of the nucleus containing the nucleolus has remained in the mother cell and is disintegrating at this stage. (g) further elongation of the spindle. One end of the spindle (within a daughter nucleus) is now located in the mother cell. (h) reconstruction of intact daughter nuclei.

Figs. 394–8 *(394–8 Glutaraldehyde–osmium tetroxide fixation.) From* MCCULLY, E. K. and ROBINOW, C. F. (1972). *J. Cell Sci.,* **10**, 857. *398 Modified from* MCCULLY, E. K. and ROBINOW, C. F. (1972). *J. Cell Sci.,* **10**, 857.

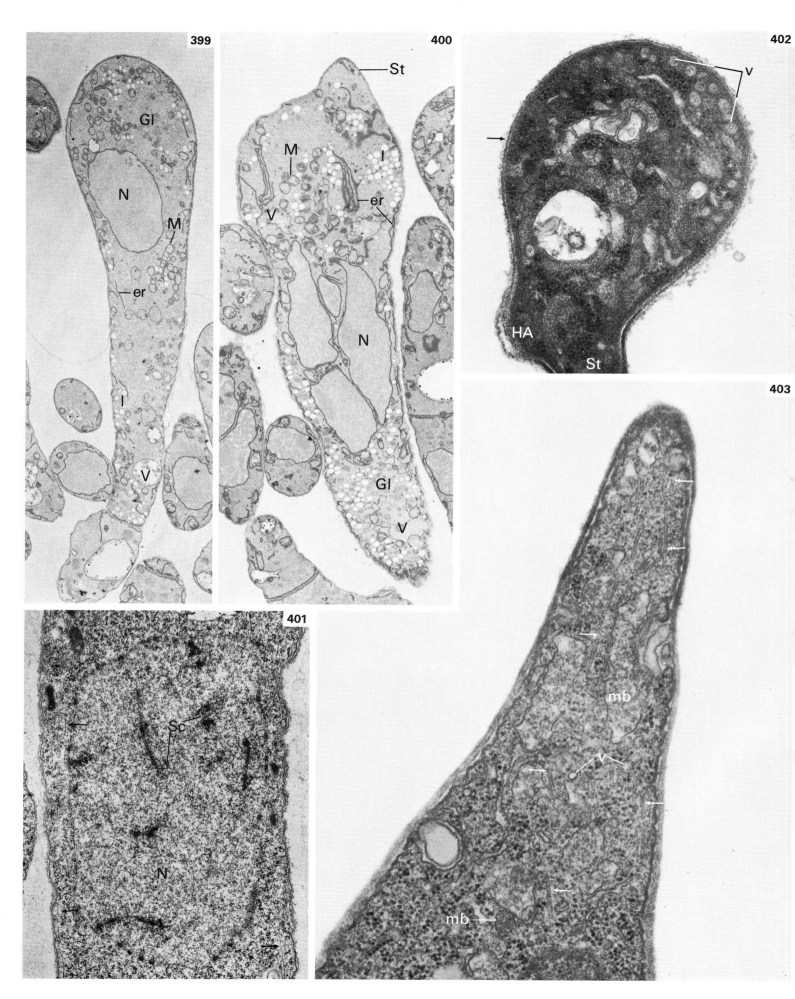

B. Reproductive Structures

Basidium development and basidiospore initiation

The young basidium resembles a hyphal apex; however, a regular series of cytoplasmic and nuclear changes occur during basidial development that are absent in hyphal development. At the time of the first meiotic prophase, the basidium of *Boletus rubinellus* Peck contains few vacuoles and limited amounts of storage products; for example, glycogen and lipid droplets (fig. 399). Larger vacuoles may be present in prefusion basidia. Rough endoplasmic reticulum is scattered in the cytoplasm and may form layers of cisternae at the cell apex. During second meiotic interphase the **sterigmata** form on the basidium and large amounts of glycogen and especially lipid droplets become evident (fig. 400). Large amounts of glycogen also accumulate at this time in *Schizophyllum commune* Fr. but lipid droplets are normally absent throughout basidial development. Endoplasmic reticulum is distributed along the cell periphery in *B. rubinellus* and forms layers of cisternae near the basidial apex (fig. 400). Vacuoles begin to enlarge and eventually come to occupy most of the space in the mature basidium.

Nuclei are situated in particular regions of the basidium at different stages in basidial development. At prefusion and during the first meiotic prophase they are found in the central region of the basidium (fig. 399). Meiotic prophase I chromosomes pair and synaptonemal complexes (structures which are characteristic of this stage of meiosis; see Section II, p. 149) are formed (fig. 401). In many Basidiomycotina, nuclei then move to the basidial apex where division occurs. They then return to the middle of the basidium until spores form (fig. 400), after which they migrate back to the apex and enter the spores.

The movement of the nuclei is accompanied by changes in the distribution of cytoplasmic microtubules. In *B. rubinellus* the predominant orientation of the microtubules at prefusion and first meiotic prophase is parallel to the long axis of the basidium (fig. 401). They appear to be absent from the cytoplasm during the first meiotic division when the spindle apparatus forms but resume their longitudinal orientation in the basidium during the meiotic interphases. Microtubules are also longitudinally orientated in the developing sterigmata of *B. rubinellus* and *Coprinus cinereus* (Schaeff. ex Fr.) S. F. Gray (fig. 403) and may supply a **cytoskeleton** for the outgrowth of the sterigma from the basidial apex. In addition, microtubules may be involved in nuclear and cytoplasmic movements within the basidium and growing spores.

The growth of the sterigma and basidiospore involves vesicles which accumulate at the growing margin (fig. 402). These vesicles are presumably derived from the Golgi apparatus and probably carry wall components. The young basidiospore is surrounded by thin wall layers which are continuous with those of the sterigma (fig. 402).

Additional reading

MCLAUGHLIN, D. J. (1971). Centrosomes and microtubules during meiosis in the mushroom *Boletus rubinellus*. *J. Cell Biol.*, **50**, 737.
MCLAUGHLIN, D. J. (1973). Ultrastructure of sterigma growth and basidiospore formation in *Coprinus* and *Boletus*. *Can. J. Bot.*, **51**, 145.
WELLS, K. (1965). Ultrastructural features of developing and mature basidia and basidiospores of *Schizophyllum commune*. *Mycologia*, **57**, 236.

Fig. 399

Basidium of *Boletus rubinellus* Peck at prophase I of meiosis. A large fusion nucleus (N) is situated near the apical end of the basidium. Few vacuoles (V), a limited amount of glycogen (Gl) and lipid droplets (I) are present in the cytoplasm. (er) endoplasmic reticulum. Potassium permanganate fixation. × 6,000.

Fig. 400

Basidium of *B. rubinellus* at interphase II of meiosis with a sterigma (St) at the apex. The four nuclei (N) are near the middle of the basidium, and vacuoles (V) are developing nearby. The endoplasmic reticulum (er) is extensive toward the basidial apex and is found near the wall along the length of the basidium. Note the increased concentration of lipid droplets (I) and glycogen (Gl) in the cytoplasm compared to the basidium at prophase I. (M) mitochondrion. Potassium permanganate fixation. × 6,000.

Fig. 401

Longitudinal section through a basidium of *B. rubinellus* at prophase I of meiosis. Synaptonemal complexes (Sc) are seen within the nucleus (N). In some cases only the lateral element of a complex which is not in the plane of the section is visible. Cytoplasmic microtubules (arrows) are orientated parallel to the long axis of the basidium. Glutaraldehyde–osmium tetroxide fixation. × 18,500.

Fig. 402

A young basidiospore of *Coprinus cinereus* (Schaeff. ex Fr.) S. F. Gray. The spore is growing towards the right and upwards and a zone of vesicles (v) is situated at the growing margin. The vesicles are similar to those produced by the Golgi cisternae which occur in basidia. The spore has a thin wall (arrow) similar in thickness to that of the sterigma (St). (HA) hilar appendix. Glutaraldehyde–osmium tetroxide fixation. × 52,500.

Fig. 403

Longitudinal section of a sterigma of *C. cinereus*. Numerous microtubules, some indicated by arrows, are orientated parallel to the long axis of the sterigma. The concentration of ribosomes is higher at the base of the sterigma than at the still elongating apex. Microbodies (mb) and vesicles (v) are present. Glutaraldehyde–osmium tetroxide fixation. × 52,000.

Figs. 399–403 *399, 400 Micrographs by* D. J. MCLAUGHLIN, University of Minnesota, St Paul. *401 From* MCLAUGHLIN, D. J. (1971). *J. Cell Biol.*, **50**, 737. *402, 403 From* MCLAUGHLIN, D. J. (1973). *Can. J. Bot.*, **51**, 145. Reproduced by permission of the National Research Council of Canada.

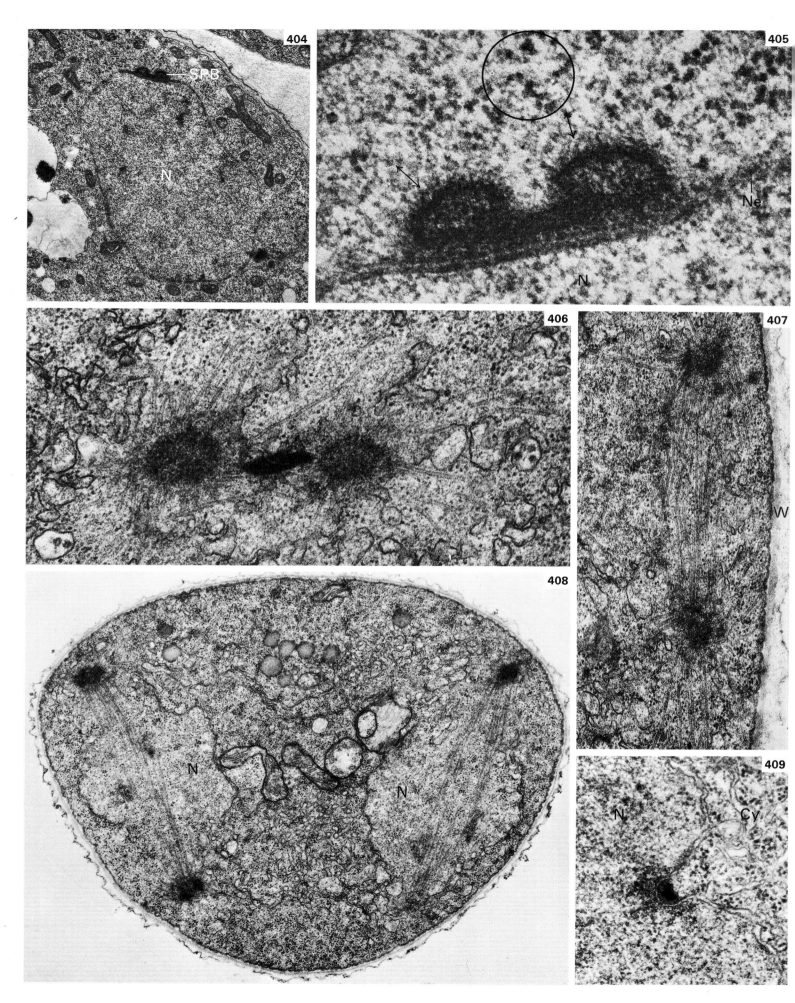

207

B. Reproductive Structures (cont.)

Meiosis

In many respects meiosis in Basidiomycotina is similar to mitosis: a **bipolar spindle pole body** is present and its globular ends separate to form the poles of the spindle. The nuclear envelope breaks down during division and chromosomes are moved to the spindle poles in conjunction with the spindle microtubules. In *Boletus rubinellus* Peck a bipolar spindle pole body, 650–700 nm long, is seen closely associated with the nuclear envelope at the first meiotic prophase (figs. 404, 405). It is composed of two flattened **globular components** joined by a **middle piece** and is surrounded by a **ribosome-free zone** where it borders the cytoplasm.

During the first meiotic prophase in *Coprinus radiatus* (Bolt) Fr. the two components of the bipolar spindle pole body separate and the middle piece appears to be lost (fig. 406). The two spindle pole bodies thus formed, act as microtubule organizing centres. They move apart in the cytoplasm and spindle microtubules develop between them (fig. 407). The spindle forms across the basidial apex. The nuclear envelope breaks down and the spindle enters the nucleus. The nucleoplasm and spindle poles are surrounded during division by what is presumed to be endoplasmic reticulum. The chromosomes become associated with **chromosomal microtubules** and move to the spindle poles. At the second meiotic division the two nuclei divide simultaneously across the apex of the basidium (fig. 408). At the second meiotic interphase the spindle pole body is found in the cytoplasm in an indentation in the nuclear envelope closely associated with dense material in the nucleus which may be chromatin (fig. 409). This is the typical position of the spindle pole body during all interphases.

Before nuclear fusion and meiosis, bipolar spindle pole bodies occur on each nucleus. Their fate at karyogamy is uncertain. It is possible that they disappear and then reform during the first meiotic prophase.

The spindle pole bodies have often been referred to as **centrioles**. However, in Basidiomycotina, centrioles with the typical substructure of nine sets of three microtubules arranged in a circle have never been found. It is therefore doubtful at present whether any homology between spindle pole bodies and true centrioles exists.

Additional reading

LERBS, V. (1971). Licht- und elektronenmikroskopische Untersuchungen an meiotischen Basidien von *Coprinus radiatus* (Bolt) Fr. *Arch. Mikrobiol.*, **77**, 308.

MCLAUGHLIN, D. J. (1971). Centrosomes and microtubules during meiosis in the mushroom *Boletus rubinellus*. *J. Cell Biol.*, **50**, 737.

Fig. 404

A bipolar spindle pole body (SPB) attached to the nuclear envelope of a fusion nucleus (N) of *Boletus rubinellus* Peck. × 14,700.

Fig. 405

A high magnification micrograph of the dumb-bell-shaped spindle pole body of *B. rubinellus*, which is situated in an indentation of the nuclear envelope (Ne). It is surrounded by a zone (double headed arrows), which contains few ribosomes (see in circle) compared to the neighbouring cytoplasm. (N) nucleus. × 102,000.

Fig. 406

A transverse section through the basidial apex of *Coprinus radiatus* (Bolt) Fr. at the time of meiotic prophase I showing the very early separation of the globular ends of the spindle pole body. Microtubules radiate into the cytoplasm from the periphery of each spindle pole body. The dark structure between the spindle pole bodies seems to be the former connection or middle piece of the original bipolar spindle pole body. Simultaneous glutaraldehyde–osmium tetroxide fixation. × 60,000.

Fig. 407

Longitudinal section through the basidium of *C. radiatus* at a later stage in the separation of the spindle poles in the cytoplasm than that shown in fig. 406. (W) apical wall of the basidium. Simultaneous glutaraldehyde–osmium tetroxide fixation. × 40,000.

Fig. 408

Transverse section through the basidial apex of *C. radiatus* during the second meiotic division. The two meiotic spindles are formed simultaneously by the development of microtubules from the periphery of the spindle pole bodies. (N) nucleus. Simultaneous glutaraldehyde–osmium tetroxide fixation. × 22,000.

Fig. 409

Cross section of a spindle pole body of *Coprinus cinereus* (Schaeff. ex Fr.) S. F. Gray at meiotic interphase II. The dense material within the nucleus adjacent to the spindle pole body may be chromatin. (N) nucleus; (Cy) cytoplasm. Glutaraldehyde–osmium tetroxide fixation. × 52,500.

Figs. 404–9 (*404, 405 Glutaraldehyde–osmium tetroxide fixation.*) *From* MCLAUGHLIN, D. J. (1971). *J. Cell Biol.*, **50**, 737. *406 Micrograph by* DR V. LERBS, Pflanzenphysiologisches. Institut der Freien Universitat, Berlin. *407, 408 From* LERBS, V. (1971). *Arch. Mikrobiol.*, **77**, 308. *409 Micrograph by* D. J. MCLAUGHLIN, University of Minnesota, St Paul.

Reproductive Structures

Meiosis

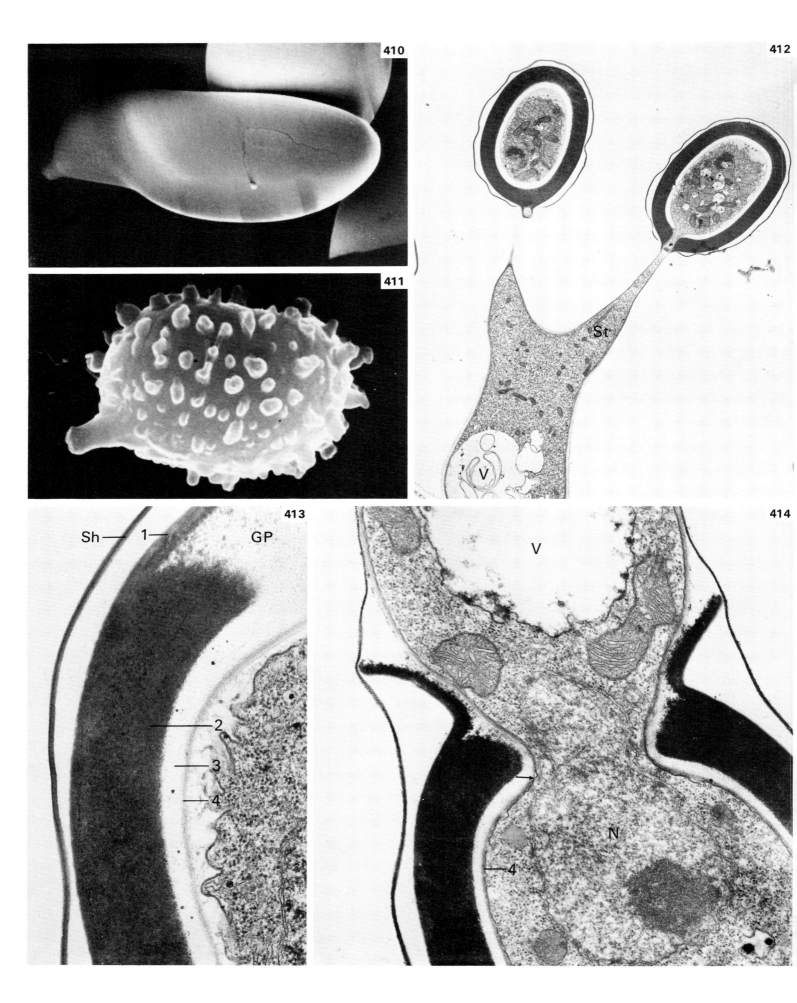

410

411

412

St

V

413

Sh — 1 — GP

2

3

4

414

V

N

4

B. Reproductive Structures (cont.)

Basidiospores and basidiospore germination

Basidiospores exhibit a wide range of surface morphologies, for example, figs. 410, 411: the scanning electron micrograph of a basidiospore of *Boletus mirabilis* (Murr.) Singer reveals a smooth surface while that of *Lactarius deceptivus* Peck shows large isolated projections. The **hilar appendix** by which the spore was attached to the **sterigma** is on the left. Scanning electron microscopy of basidiospores and transmission electron microscopy of carbon replicas, as well as suggesting new relationships of some genera and species have confirmed many family alliances in Hymenomycetes and Gasteromycetes which were previously based on light microscope evidence.

Mature basidiospores have thicker and more complex walls than the basidium (fig. 412). The development of vacuoles in the basidium may provide the hydrostatic pressure necessary to fill the basidiospore with cytoplasm. The mature basidiospore wall of *Coprinus stercorarius* (Bull. ex St. Amans) Fr. has four wall layers and a loose outer sheath (fig. 413) and is greatly thickened compared with a newly initiated spore (see fig. 402). There is no wall layer between the outer sheath and layer 1. The **germ pore** which develops at the spore apex consists mainly of wall layer 3, and wall layer 2 is largely absent except for a thin layer over the top of the germ pore. A complex terminology exists for the wall layers of basidiospores. While *C. stercorarius* has thick-walled basidiospores, some Hymenomycetes form thin-walled basidiospores which have fewer wall layers and lack germ pores. Two types of **hilum** have been demonstrated in studies of surface morphology of the hilar appendix: the **open-pore** and the **nodulose** types. Both the open-pore type of hilum and the germ pore involve modifications in the spore walls and may be structurally related.

Spore germination in *C. stercorarius* involves the breaking of the outer spore walls of the germ pore through which the germ tube emerges (fig. 414). The new hyphal wall is continuous with the innermost layer of the spore wall.

Additional reading

HEIM, R. and PERREAU, J. (1971). Étude ornementale de basidiospores au microscope électronique à balayage, in: *Scanning Electron Microscopy*. (V. H. Heywood, ed.) pp. 251–284. Academic Press, London.
HEINTZ, C. E. and NIEDERPRUEM, D. J. (1971). Ultrastructure of quiescent and germinated basidiospores and oidia of *Coprinus lagopus. Mycologia*, **63**, 745.
PEGLER, D. N. and YOUNG, T. W. K. (1971). Basidiospore morphology in the Agaricales. *Nova Hedwigia*, **35**, 1.

Fig. 410

Scanning electron micrograph of a smooth-walled basidiospore of *Boletus mirabilis* (Murr.) Singer seen in lateral view. The hilar appendix is on the left or basal end of the basidiospore. × 6,000.

Fig. 411

Scanning electron micrograph of an ornamented basidiospore of *Lactarius deceptivus* Peck seen in lateral view. The surface bears large, isolated projections. The hilar appendix is to the left. × 7,000.

Fig. 412

Mature basidiospores of *Coprinus stercorarius* (Bull. ex St. Amans) Fr. attached to sterigmata (St). The basal vacuole (V) in the basidium is in the process of enlarging and filling the spores with cytoplasm. Glutaraldehyde—osmium tetroxide fixation. × 5,800.

Fig. 413

Part of a mature basidiospore wall of *C. stercorarius*. It is composed of four wall layers which are numbered in order of their probable initiation. The basidiospore is surrounded by a sheath (Sh) and part of the apical germ pore (GP) is visible. Glutaraldehyde—osmium tetroxide fixation. × 42,900.

Fig. 414

Germinated basidiospore of *C. stercorarius*. Wall layer 4 of the spore is continuous with the wall of the germ tube (arrow). A vacuole (V) is present in the germ tube just exterior to the germ pore. (N) nucleus. Glutaraldehyde—osmium tetroxide fixation. × 29,000.

Figs. 410–14 *410 From* GRAND, L. F. and MOORE, R. T. (1971). *Can. J. Bot.*, **49**, 1259. Reproduced by permission of the National Research Council of Canada. *411 From* GRAND, L. F. and MOORE, R. T. (1970). *J. Elisha Mitchell scient. Soc.*, **86**, 106. *412–14 Micrographs by* DR M. A. ROGERS, Department of Botany and Plant Pathology, Iowa State University, Ames.

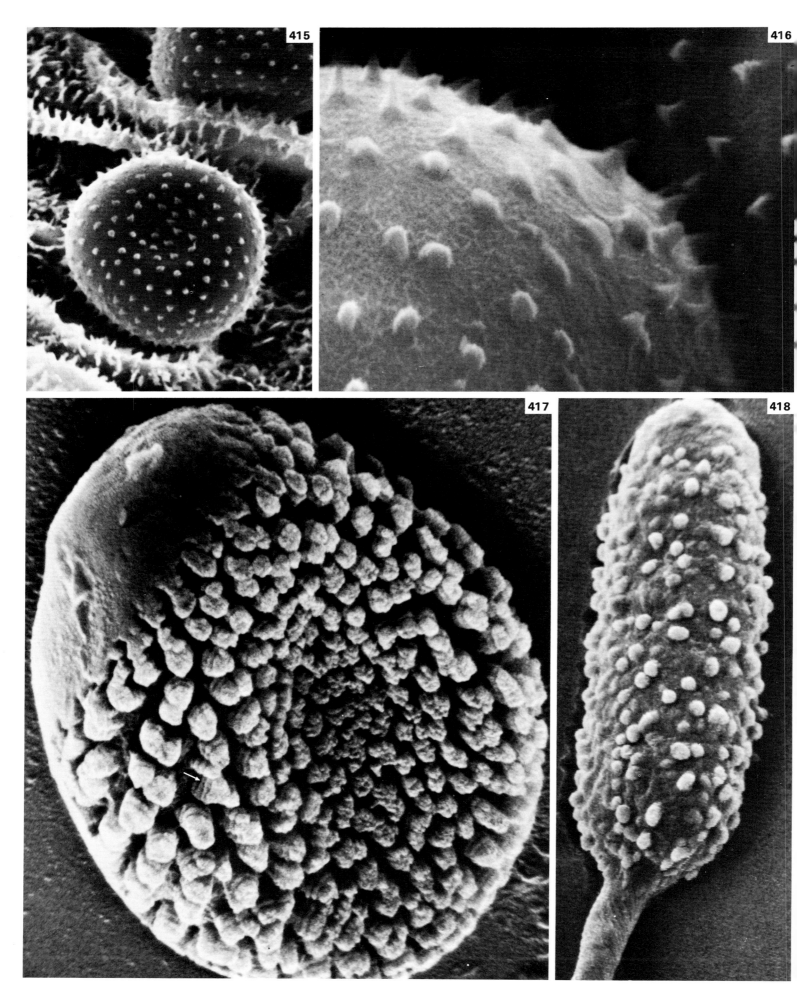

B. Reproductive Structures (cont.)

Rust spores

Surface morphology of rust spores and those of other Teliomycetes has been extensively used for taxonomic purposes. Descriptions based on light microscopy may be inaccurate or inadequate. The scanning electron microscope is now providing a clearer understanding of the topology of the spore surface.

Aeciospores, urediniospores and teliospores of rusts (Uredinales) show distinctive morphologies. In addition the manner in which these surface structures develop appears to be characteristic for each type of spore. Urediniospores of *Melampsora lini* (Ehrenb.) Lév. are covered with spines (figs. 415, 416), which are initiated within the spore wall next to the plasma membrane and seem to be of different chemical composition from the wall. Within the cytoplasm the endoplasmic reticulum is closely associated with the developing spines. As the spines enlarge and mature they become separated from the plasma membrane by centripetally formed wall material. At maturity the spines project from the outer surface of the uredinio- spore with only their bases embedded in the wall. Thus, the spines are initiated within the wall next to the cytoplasm but come to rest on the outer surface of the wall.

The development of the ornamentation on aeciospores is different from that of urediniospores. In *Cronartium quercuum* (Berk.) Miy. ex Shir. the spore surface consists of a series of ringed projections and a smooth area to which some of the projections are connected (fig. 417). The ornamentation is presumably formed by the cracking of the outer wall of the aeciospore to form projections as the young spore expands. This concept of the development of the surface morphology is supported by the facts that the smooth area on the spore has the same height as the projections and is continuous with them, and that pro- jections adjacent to the smooth area appear to fit together.

Teliospores of *Phragmidium occidentale* Arth. bear irregular projections on their surface (fig. 418). The pro- jections on teliospores apparently arise either by localized outgrowths of the spore surface which are of similar com- position to that of the adjacent spore wall, or by re- absorption of part of the initially smooth spore wall.

Additional reading

GRAND, L. F. and MOORE, R. T. (1970). Ultracytotaxonomy of basidiomycetes 1. Scanning electron microscopy of spores. *J. Elisha Mitchell scient. Soc.*, **86**, 106.

LITTLEFIELD, L. J. and BRACKER, C. E. (1971). Ultrastructure and development of urediospore ornamentation in *Melampsora lini. Can. J. Bot.*, **49**, 2067.

Fig. 415

Scanning electron micrograph of a urediniospore of *Melampsora lini* (Ehrenb.) Lév. The spore at the centre of the figure is seen in end view and is covered with electron-transparent spines arranged in a nearly helical pattern. × 3,000.

Fig. 416

A high magnification micrograph of the urediniospore surface reveals the sharp-pointed spines and the uneven reticulated pellicle on the surface between them. × 14,000.

Fig. 417

A scanning electron micrograph of an aeciospore of *Cronartium quercuum* (Berk.) Miy. ex Shir. The aeciospore surface is covered with projections on some of which characteristic ringed markings (arrow) can be seen. At the upper end of the spore the projections are continuous with a smooth area. × 9,000.

Fig. 418

A scanning electron micrograph of a stalked teliospore of *Phragmidium occidentale* Arth. Irregular wart-like projections are distributed over the spore surface. The stalk at the lower end of the spore has a wrinkled, wart-free surface. × 3,000.

Figs. 415–18 *415, 416 From* LITTLEFIELD, L. J. (1971). *J. Microscopie*, **10**, 225. *417 From* GRAND, L. F. and MOORE, R. T. (1972). *Can. J. Bot.*, **50**, 1741. Reproduced by permission of the National Research Council of Canada. *418 Micrograph by* DRS L. F. GRAND and R. T. MOORE, North Carolina State University, Raleigh, N.C.

GLOSSARY

Apical body. See Spitzenkörper.

Astral ray microtubules. Cytoplasmic microtubules which are associated with spindle pole bodies and which may be involved in ascospore delimitation.

Axoneme. The '9 + 2' microtubular core of a flagellum. (See Flagellum.)

Cell wall. Predominantly polysaccharide layer which encloses most plant cells.

Centriole. Organelle composed of an approximately 0·2 μm long cylinder of nine triple microtubules. Typically found in pairs adjacent to the nucleus and at the poles of dividing nuclei.

Chromosomal microtubule. A microtubule of the spindle which extends from the spindle pole to a kinetochore on a chromosome.

Collar. A deposit of material upon the host cell wall around the region of a penetrating haustorium.

Continuous microtubule. A microtubule of the spindle which is not attached to chromosomes and may extend from one spindle pole to the other.

Cytoplasm. Material in which organelles and ribosomes are suspended inside the cell, the 'ground substance' of a cell.

Dictyosome. A stack of plate-like membranous cisternae. The side of the stack to which fresh cisternae are added is termed the *forming face* and the opposite face from which vesicles are produced is called the *secretory face*. The Golgi apparatus of a cell may consist of one or many dictyosomes, or a number of related cisternae.

Elementary particles (Oxysomes). Stalked spheres seen on the inner mitochondrial membrane after negative staining; thought to contain some of the enzymes involved in the electron transport system.

Endoplasmic reticulum. Network of membranes forming flattened fenestrated sheetlike chambers or tubules (cisternae) which permeate the cytoplasm. May have ribosomes attached to the cytoplasmic side of the membranes (*rough* e.r.) or may not (*smooth* e.r.)

Flagellum. Whiplike organelle used for propulsion. Has a membranous sheath and an axoneme or core composed of a cylinder of nine double microtubules enclosing two central single microtubules ('9 + 2' arrangement). May be of two types, *tinsel* (or *flimmer*), with characteristic hairs attached to the sheath, or *whiplash* which lacks these hairs.

Flimmer hair (Mastigoneme). Lateral hair or projection of the tinsel-type flagellum.

Golgi cisternae. Flattened membrane bound sacs often with dilated or vesicular periphery from which vesicles are produced.

Haustorium. A specialized organ which is formed inside a living host cell as a branch of an extracellular or intercellular hypha and which is probably involved in the interchange of substances between host and fungus.

Kinetochore (Centromere). Subunit of a chromosome which functions in chromosomal movement during division and to which the chromosomal microtubules of the spindle are attached.

Kinetosome (Basal Body). Cylinder of nine triple microtubules which form the base of a flagellum. Frequently formed by elongation of a centriole.

Lomasome. Group of membranous tubules or vesicles lying between the plasma membrane and the cell wall. Vesicles may be embedded in the cell wall.

Microbody. Small, usually spherical body which is bound by a single membrane. It has a characteristically staining lumen which often contains an osmiophilic crystal or amorphous body. This is a descriptive term for bodies which may subsequently be defined biochemically as **peroxisomes, glyoxysomes** or **lysosomes**.

Microtubule. Characteristic 20–25 nm diameter apparently rigid proteinaceous tubule found in the cytoplasm, in nuclear division spindles and in flagellar axonemes.

Middle piece. The central part, which connects the two ends of a bipolar spindle pole body. Typically found in Basidiomycotina.

Mitochondrion. Organelle in which part of the respiratory system is located. Bounded by a smooth outer membrane and a convoluted inner membrane which forms finger-like or lamellar invaginations termed *cristae.*

Negative staining. Technique whereby material is surrounded by an electron-opaque stain which outlines the structure of the specimen.

Nucleoplasm. Material within the nuclear envelope in which nuclear ribosomes, the nucleolus and chromosomes are suspended.

Osmiophilic. Binds osmium more strongly than adjacent regions of tissue and thus becomes more electron-opaque and consequently appears dark in electron micrographs.

Perimitochondrial space. The space between the outer and inner membrane of the mitochondrial envelope. Probably filled with aqueous liquid.

Plaque bridge. An electron-opaque rod-like structure which connects the daughter halves of the replicating spindle pole body in *Saccharomyces cerevisiae.* May be homologous to the **Middle Piece** of Basidiomycotina.

Plasma membrane (Plasmalemma). Membrane which forms the boundary between the cytoplasm and the cell wall or external environment.

Plugging precursor. Material present in the septal

Glossary continued

pore apparatus which occludes the pore when the cell is disrupted or dies.

Ribosome. A 15–25 nm mass of nucleoprotein. Ribosomes may be found in the cytoplasm or associated with endoplasmic reticulum and the nuclear envelope. They may aggregate into groups to form **Polyribosomes** which function in protein synthesis.

Septal plug. An occlusion in a septal pore.

Septal pore apparatus. The septal pore with its associated structures.

Septal pore cap (Parenthesome). A hemispherical structure related to the endoplasmic reticulum which encloses the septal pore swelling.

Septal pore swelling (Dolipore). A doughnut-shaped enlargement bordering the pore in the septum.

Sheath (Encapsulation). An amorphous region between the haustorial wall and host plasma membrane.

Spindle. A group of microtubules which extend between two polar regions and which function in the separation of chromosomes during nuclear division. Typically composed of **Chromosomal microtubules** and **Continuous microtubules**.

Spindle pole body. A structure of variable size and morphology associated with nuclei, spindles and microtubules in non flagellate fungi.

Spitzenkörper. A region of granular and/or vesicular cytoplasm at the apex of a growing hypha.

Synaptonemal complex. A structure which holds the two homologous chromosomes of a bivalent in precise alignment along their entire length. Composed of **lateral** and **central** components.

Tonoplast (Vacuolar Membrane). The membrane which forms the boundary between the cytoplasm and the vacuole.

Unit membrane. Trilamellar lipid-protein sheet-like structure seen as a non-osmiophilic (light) layer sandwiched between two osmiophilic (dark) layers when viewed in transverse section in the electron microscope.

Vacuole. Membrane bound cavity within the cytoplasm which typically contains aqueous nonprotoplasmic liquids or storage products.

Vesicle. A small (approx. 0·1–0·5 μm), membrane-bound, usually spherical, body which may be associated with any of the several membrane systems of the cell. May function in transport of substances within the cell (Golgi vesicles, wall vesicles, lomasome vesicles).

Wall vesicles. Small single membrane-bound vesicles occurring in regions of cell wall synthesis.

AUTHOR INDEX

SUBJECT INDEX

Page numbers in *italic* indicate illustrations of genera and species

Subject Index continued